Our States Have *Crazy* Shapes

Lynn Garthwaite

Also by Lynn Garthwaite

Our States Have *Crazy* Shapes

Panhandles, Bootheels, Knobs and Points

Lynn Garthwaite

Cover illustration by Scott Spinks

Blue Spectrum Books
Bloomington, MN

Published by Blue Spectrum Books
8425 Quinn Avenue S, Bloomington, MN 55437

Cover illustration by Scott Spinks:
www.scottspinks.com

Cover and Interior Design by FuzionPrint
www.fuzionprint.com

Maps by Beth M. Anderson: www.plum-creek.com

Library of Congress Control Number:
2016903980

ISBN – Paperback:
978-0-9973967-0-6

To all of those curious people who scan newspapers and books, searching for quirky tidbits that make you say— "That is so cool!"

Table of Contents

Acknowledgments

This book would not exist if it hadn't been for the extensive and fascinating research accomplished by **Mark Stein** for his book *How the States Got Their Shapes, (HarperCollins, 2008)*. I bought his book to find the answer to a question I've wondered about: why is the Upper Peninsula part of Michigan and not Wisconsin?

Since not everyone is inclined to delve into a textbook on the subject, I set out to write a more anecdotal version. My emphasis is on the quirkiest parts of every state's shape and not intended as a comprehensive treatise on all of the aspects that went on behind the scenes. It is my wish that you find this an excellent way to *accidentally* learn a little bit about American history.

While Stein's book was my primary source, several other wonderful books and materials led to additional information on the fascinating stories behind how our states got their crazy shapes. Note the short bibliography in the back pages. But if reading this book raises your curiosity about some of the finer details and all of the perplexing and confounding events that merged to give us this jigsaw puzzle of a map, you're going to want to read Stein's book.

Thank you to my amazing editors, Claudette Hegel and Connie Anderson. You helped make this an intelligible final product.

Thank you to my mapmaker, Beth M. Anderson, who also designed my website:
www.bluespectrumbooks.com

Thank you to my printer, Ann Aubitz with FuzionPrint
www.fuzionprint.com

And thank you to my cover illustrator, Scott Spinks.
www.scottspinks.com

Every single one of you saved my patootie!

Acknowledgment for the photos of survey demarcation on page 17:

Bryan Cloutier
Engineering Technician
International Boundary Commission
Central Regional Field Office
Thief River Falls, MN

Lynn Garthwaite

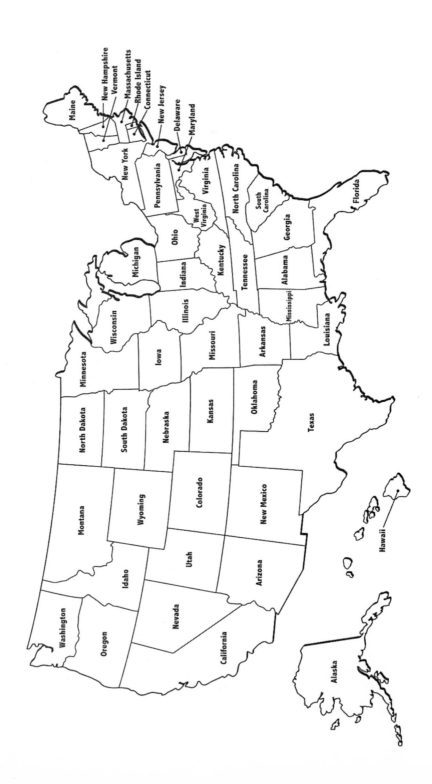

Introduction: Dividing the Land

The original mapmakers might have been a touch crazy. The fifty United States are made up of such odd and irregular shapes and sizes, it's as if the planners just randomly chose landmasses and drew borders around them. How is it possible that we have a state as large as Texas and as small as Rhode Island in the same country? And why are some states so irregular in shape while others are plain rectangles?

America was built in stages, beginning with the thirteen original colonies. With westward expansion, the United States formed The Northwest Territory that later split into five states. As settlers moved across the Mississippi and the country bought the land known as the Louisiana Purchase, the size of the Union grew immensely.

In some cases, a state's border seems obvious because the meandering course of a major river defined it, but how did Michigan end up with the Upper Peninsula when Wisconsin is the state attached to it by land? Why does Delaware have a northern border that is a perfect arc? Who decided where North Dakota stopped and South Dakota started? How did we end up with a nation full of panhandles, bootheels, knobs and points?

The answers are sometimes startling and frequently explained by the political squabbles in treaty settlements, land grants, and the spoils of war. Some of the borders standing today were actually the result of errors by surveyors who were using 18th-century tools without benefit of modern roving satellites to mark the lines.

But before we examine exactly *where* the borders were drawn, we need to understand *how* men physically carved the borders that often bisected large areas of wilderness and lakes. Drawing lines on a map is easy, but we needed a special group of people charged with the task of drawing those lines on the ground.

Surveying with the purpose of defining borders actually dates as far back as 5,000 years to the Middle East and Egypt. Rulers knew the value of establishing property lines because firm boundaries allowed for a way to extract revenue from the sale of land, and to continue collecting taxes from new landowners. Delineating borders also gives a central government clearly defined spheres of influence.

The situation was much the same in our newly established United States. The fledgling American government had accumulated massive debts from various wars with both the British and the Spanish, and now was greatly in need of capital to finance the work of a nation. As new settlers arrived, they were claiming land without giving any compensation to the government, so surveyors were dispatched to establish boundaries.

Among the first to work in this capacity were the now-famous Charles Mason and Jeremiah Dixon. Like all other surveyors at the time, Mason and Dixon were astronomers who used a variety of telescopes and star charts to do their job. In order to define on land the border that had only been described on paper using longitudes and latitudes, the skill of the astronomer ultimately divided the land.

Astronomers like Mason and Dixon received specific guides from the king of England with which to mark the borders. For example, as defined by the king, the land to be known as New England was described as:

> "From Fourty degrees of Northerly Latitude, from the Equnoctiall Line, to Fourty-eight degrees of the said Northerly Latitude."

Mason and Dixon's specific instruction was to work on the first surveying task, the southern border of Pennsylvania. Because the city of Philadelphia had already been established and had become the site of the first United States capital by that time, the state of Pennsylvania was deemed a crucial place to begin. Congress determined the southern border would be a straight east/west line exactly 15 miles south of Philadelphia, so Mason and Dixon gathered their team to mark that line.

The project took them four years, from 1763–1767. In addition to the two astronomers, the team consisted of about two dozen assistants, a team of packhorses to carry all the equipment, and 100 ax men to physically sculpt the border through the forests.

The astronomers positioned themselves on high points along the way and spent many evenings observing the positions of the stars in the sky. They consulted their charts and used their knowledge of the heavens to determine exactly where they were with regard to the lines of latitude and longitude. When they agreed they were in the correct place, their assistants pounded in a marker, and the ax men followed behind, chopping down trees to form a line of demarcation. After months of exhausting labor, the group looked behind them and saw with satisfaction a straight line cleared in the wilderness to mark the border.

The process was excruciatingly slow. In four years Mason and Dixon were able to establish just the one task they were assigned, the southern border of Pennsylvania, along with the western and southern borders of Delaware. Their finished product became known as the Mason-Dixon line.

These pictures, taken much more recently, show the demarcated border between the U.S. and Canada. Although more modern equipment created these lines, we get a peek at what our earliest surveyors accomplished. (Photos provided by The International Boundary Commission):

As the years passed, Congress hired more astronomers to survey other state borders, and gradually the equipment improved. One of the best known of these new astronomers was American Andrew Ellicott, a clockmaker who parlayed his skill with timepieces into making extremely advanced telescopes for the time. Mostly self-taught, Ellicott quickly became one of the highest regarded surveyors of the era and later mapped out the land along the Potomac when the nation's capital moved from Philadelphia to its current location in Washington, D.C.

One of the early challenges Ellicott and other astronomers had to overcome was the difficulty of drawing a straight line across the rounded surface of the globe. Poised on a hill in the middle of the wilderness, surveyors like Ellicott needed to account for the curvature of the earth when calculating the correct direction of the line through the forests. Moving so many people and animals through uncharted areas was frequently dangerous since they passed through mountains with narrow ledges, river crossings, and foliage growing so thick they had to cut through it in order to move forward. Falls, broken bones, and the loss of important pack animals were all part of the surveyors' challenges.

Unforeseen disasters on the other side of the world also interrupted their process. In 1783, a volcano erupted in Iceland and the resultant ash in the air blocked the view of the stars as far south as North America, delaying Ellicott's progress for several weeks.

In spite of the hardships, no one doubted the importance of the surveying process. Without the distinction of borders, state laws carried no weight with the new settlers. Border disputes were a constant hurdle in those early days of settlement, and for law enforcement officers, state boundaries had become crucial to keeping the peace.

Those early years of grueling work in the wilderness heavily burdened the surveyors and their teams of workers, but the result was a nation of states, counties, cities and towns that survive to this day with an order and clarity the frontier did not have. In spite of handmade equipment, antiquated by today's standards, their accuracy was remarkable. When Ellicott and his team were asked to continue surveying Pennsylvania in 1784, they drove in a large, white squared post to mark the southwest corner. Modern GPS measurements have indicated the post, which remains to this day, sits a mere 23 feet off the correct mark.

Hey – This Stuff Is Interesting!

*T*hese are some of the sidebar tidbits that don't fit into a single state's chapter, but rather apply to many areas. Reading these chapters ahead of time will help make sense of some of the decisions made as each state was formed.

Squiggly States in the East, Boxy States in the West

When you look at a map of the United States you can't miss the fact that the states in the east appear to have been cut by a completely different jigsaw than the states in the west. On the east coast, states tend to be small, with a mix of odd zigzag shapes. West of the Mississippi River, the states are noticeably larger, with borders made of straight lines. Some even have 90° corners. What happened?

In a word: railroads.

When territories claimed their borders in the early days of colonization, the primary means of moving people and commerce were rivers. A colony's economic health hinged on having access to the many rivers and bays that crisscrossed the east coast. Without river access, a colony or state would have to depend on its neighbors to provide important routes

21

to move goods for sale. As a result, the course of water primarily determined borders in our early years. When a river divided two states, both sides had ample access.

By the time large numbers of settlers moved west, railroads took over as the main means of transportation. Instead of sending harvested crops up and down the rivers, farmers loaded them on trains to move from town to town. A business that produced building materials and tools used the railroad to deliver their product to stores. With lessened importance on access to rivers, borders could be drawn along straight lines.

States That Didn't Make the Final Cut: Franklin and Deseret

For four years, settlers in the northeast corner of modern-day Tennessee proudly proclaimed themselves residents of a brand-new state they called *Franklin*. At that time, the 1780s, our new post-revolution Congress had not fully settled the conditions under which a territory could apply for statehood. Residents, eager to create their own entity, simply forged ahead and declared themselves a state.

In those four years, Franklin established its own legislature, elected a governor, and wrote a state constitution. But descriptions of where various territories started and stopped were a bit slippery in those days, and the larger territory of Tennessee claimed Franklin had nabbed land belonging to Tennessee. A bloody battle erupted and in 1788, Franklin yielded. In addition to the difficulty of fighting a border battle, their coffers were bare. Tennessee ultimately absorbed the state of Franklin.

But pride in one's past doesn't die easily, and if you travel to that part of Tennessee today you may find yourself doing business with a number of companies that use "State of Franklin" in their name.

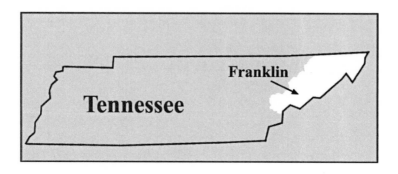

The State of Franklin had a four-year life before Tennessee claimed the land. The State of Deseret lasted only two years, and never actually officially became a state.

The word "deseret" means "honeybee" in the Book of Mormon. When early Mormon pioneers settled in the Salt Lake Valley, they envisioned a territory that supported their followers' beliefs. In 1849, Latter Day Saints President Brigham Young initially drew up papers to propose a territory, but he quickly changed it to a petition for statehood when he saw California was applying to be admitted to the Union. California and Brigham Young wanted some of the same territory.

Proposed state of Deseret

Church elders drafted a constitution and sent it by courier to Washington, D.C. The federal government responded with a proposal to combine California and Deseret as one state, but it received no support from either of the entities involved. The next year, 1850, Congress instead created the Utah Territory and named Brigham Young Utah's first territorial governor. In 1851, the proponents of the State of Deseret voted to cease pursuit of the state, although behind the scenes many continued to keep pushing the idea.

Eventually the innovation of the railroad brought large numbers of non-Mormon settlers to the area, and the notion of a Mormon state faded. The federal government never recognized The State of Deseret but the short-lived assembly made laws, formed a militia, formed counties, appointed judges and performed other governmental duties until the territorial government replaced Deseret.

A River Runs Through It,

and That's a Problem

Rivers have been the foundation for boundaries between many states, but problems arise when those rivers change course over time. Because of both slow erosion and fast flooding, a river border can cause problems for residents who may suddenly find themselves citizens of the neighboring state they once viewed across a body of water.

The Missouri River has been the spotlight for several cases that have gone to the Supreme Court because it is an unpredictable, fast-moving river that has changed course over time. The stretches of the Missouri that separate Nebraska from South Dakota and Nebraska from Iowa have resulted in more than one trip to the Supreme Court, but the issues remain unresolved. Residents of both sides of the river have cause for concern.

Courts make a distinction regarding how quickly the river's flow changed. If it happened very slowly, so slowly you can't actually see it happening (i.e. gradual erosion), it's called *accretion,* and the law declares the boundaries change as the waterway changes. If the change happens overnight, like from a flood, they call it *avulsion* and the boundaries stay the same even though the water now takes a different course.

One of the problems, in many cases, lies in the inability to figure out exactly where the boundaries were in the first place because, for large rivers and the passage of hundreds of years, both accretion and avulsion have probably happened multiple times. Nebraska and its neighbors are engaged in an ongoing legal battle over where the boundaries truly lie.

A similar debate is being waged on the Texas/Oklahoma border where the Red River changed course over time. As in the example above, the definitions of *accretion* and *avulsion* play a large part as the Bureau of Land Management has seized properties that, in many cases, have been in families for generations. Legal interpretations of the true location of that border are at the forefront of this battle.

See also Chapter 21 on Illinois to see how that state's capital city became detached from Illinois because of the changing course of the Mississippi River.

Water Fights, But of a Different Kind

The changing flow of rivers isn't the only reason borders remain in dispute. Errors in surveying, disagreements about markers, and disputes about which state is entitled to certain tracts of land can all force state governments to pursue judicial relief long after the issue is supposedly settled. As recently as 1998, New York and New Jersey squabbled over which state owned Ellis Island, and a Supreme Court ruling eventually granted approximately 90% of the island to New Jersey.

The Court's ruling referenced an 1833 decree in which New York and New Jersey settled a land dispute over ownership of the Hudson River. In that unusual decision, mediators awarded New York the surface of the river, and granted New Jersey some docking areas and the land *under the water*. When the debate over ownership of Ellis Island resurfaced 160 years later, the Supreme Court ruled New Jersey owned the part of the island that had been enlarged back in 1892. At that time, incoming immigrants were arriving in such large numbers more space was needed to process them. That island expansion, larger than the original Ellis Island, had been created from land dredged from the *bottom* of the Hudson River. The yellowed 1833 agreement made it clear the land belonged to New Jersey.

Tennessee and Georgia are engaged in a dispute over rights to the Tennessee River about a mile north of the border that separates the two states. When crews arrived to survey that border in 1818, they made an error that plays a role in the rift today. Instead of locating the border at the prescribed 35th parallel, surveyors inadvertently marked the line of demarcation one mile south.

Now, almost 200 years later, this mistake has taken on important ramifications because dry Georgia needs water from the Tennessee River. Georgia's argument is that if the border had been surveyed correctly, they would have clear ownership of and access to that part of the river. Tennessee argues that a border that has been considered accepted and legal for 200 years should not be moved simply because Georgia has not adequately handled its own water use. As of this writing, the debate has not been settled.

Jefferson's Ideal of Equal States

Thomas Jefferson had a vision for the United States long before the country expanded into the nation the world knows today. The British heavily influenced the sizes and shapes of the states as our nation began, but after the American Revolution, 1765-1783, Jefferson asserted that, going forward, all states should be equal, and it would begin with size.

States of a similar size would enjoy balanced representation in government as well as an opportunity for similar measures of natural resources and populations. With an eye toward expanding as far as the Mississippi River, Jefferson produced this grid of states, with an assortment of mostly Greek names, to represent his concept.

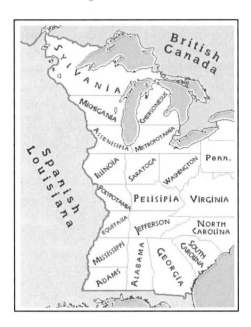

As much as possible, later Congresses that were charged with the task of assigning borders, applied Jefferson's ideal to new states as they formed. The exceptions are obvious on maps, but chapters on each of those states explain why Congress did not divide those territories the same way they divided the rest of the country.

Jefferson's ideal explains why states like North Dakota, South Dakota, Kansas and Nebraska, are very nearly the same size. All have the same three degrees in height, and very close to the same east/west measurements. The states of Montana, Wyoming and Colorado all have four degrees of height.

Colorado, Wyoming, Oregon, Washington, North Dakota, South Dakota and Nebraska are all seven degrees wide.

Looking at the states formed from the divided Carolina Territory, you see Mississippi, Alabama and Georgia revealed as essentially the same size.

When Congress divided The Northwest Territory, they created Ohio, Michigan, Indiana, Illinois and Wisconsin, all very close to the same size.

Congress' decisions on where to place borders were predominately influenced by the following thought: How will this border affect the way the remaining space can be divided in the future?

Thomas Jefferson would have approved.

There is a story behind the shapes of our states. The first of these stories is Delaware.

Chapter 1:
Delaware

First state, admitted December 7, 1787

D elaware has the distinction of being the first of
the original thirteen colonies to ratify the
Constitution of the United States and become
a state. The early settlers were Dutch, and in fact the
state was named after Lord De La Warr who was a
Dutch explorer and an early governor of the Virginia
colony.

The name "Delaware" stuck for the state, the bay, the
river and an indigenous people.

Delaware probably felt a lot of love in the beginning
because other territories wanted her. Pennsylvania
wanted to absorb the territory so their access to
Delaware Bay would not be blocked. Maryland was
pretty convinced that Delaware already was part of
their territory because of the way their royal charter
read. Eventually King Charles II of England wrote that

Maryland misunderstood the terms of the charter and he took the initiative to make decisions about how Delaware's borders would shape up.

One of Delaware's most distinctive features is a perfect arc for its northern border. Most other colonial borders were either formed by meandering rivers, straight lines from point A to point B, or were etched by coastal waters. But somebody somewhere made the decision to turn the northern border of Delaware into a smooth arc. What's up with that?

King Charles II came up with the idea to create the unusual arc on the northern border and he declared the border be set as a 12-mile radius with a center point at the Old Dutch Church. Honoring a Dutch church was an interesting decision by an English king all by itself, especially when he had kicked the Dutch out in 1674, but that anomaly straightened itself out later when the king changed the center point of that arc to the courthouse. From there, they could make some fine-point adjustments and eliminate a pesky tiny wedge of land created from the previous position of the arc. Even with the change of starting point, the arc still lands just east of the prescribed western border, so commissioners called an audible to create a snippet of a line to get the two points to meet.

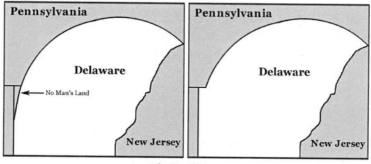

Delaware and its Arc

If Maryland wasn't already ticked enough about losing the land they thought was supposed to be theirs, they had another jab in store for them, although this one self-inflicted. Human error was the cause, and, as it turned out, human error would be the cause of bad map drawing all across the country. When Lord Baltimore, the colonial governor of Maryland, ratified the map to show the boundaries of the two territories, an incorrect reference point foiled his plans. His advisors confused Cape Henlopen with Fenwick Island, and as a result the Maryland contingency drew the southern border of Delaware about 25 miles farther south than everyone expected, cutting well into Maryland's rightful territory. Maryland asked for a do-over, but the border remains to this day.

Delaware has another unusual feature: bits and pieces of it *are actually sitting* on the New Jersey shore.

In the early 20th century, workers dredged parts of the Delaware River to create deeper shipping lanes, and they dumped the sediment pulled from the river onto the New Jersey shore. That material dredged out of the river came from Delaware property, so technically those tiny parts sitting on the New Jersey shore are part of Delaware. However, the two sides have declined to fight over the property because the area is not fit for building, but that explains the tiny speck of Delaware that appears on New Jersey maps.

Chapter 2:
Pennsylvania

Second state, admitted December 12, 1787

Villiam Penn was an English real estate entrepreneur, a philosopher and a Quaker. In 1681 King Charles II handed over present-day Pennsylvania to William Penn to pay off an old debt. The payment probably sounded like a good deal, but Penn had to accept it sight unseen because the territory was across the Atlantic Ocean from where he lived in England. Weeks later, Penn sailed to the new country to see if he got his money's worth.

Penn's plan was to name the beautiful new colony "Sylvania," which meant "woods," but the king insisted on the name "Pennsylvania." The humble William Penn, influenced by his Quaker upbringing, felt embarrassed by the name and worried people would think he named it after himself. But one learns early in life that it doesn't pay to argue with the king, so the name stood.

The original southern border of Pennsylvania was described as 40° latitude. What should have been a simple line turned out to be the subject of great dispute and the impetus of a historic boundary.

Without realizing it from across the ocean, the king's prescription of the 40th parallel as a southern border for the Pennsylvania territory actually cut right through the middle of Philadelphia. By the time legislators got serious about surveying the borders, Philadelphia was a thriving city and also served as the site of the Federal Capital. Having the border of Pennsylvania bisect its major city was an obvious issue, but changing that boundary meant Maryland was going to lose a big chunk of land. Almost 100 years of bickering ended when Congress decided to locate Pennsylvania's southern border exactly 15 miles south of Philadelphia. Maryland lost the argument, not for the first or last time in their formation.

English surveyors Charles Mason and Jeremiah Dixon surveyed that renegotiated border in what was to become the famous Mason-Dixon line. The introductory chapter on how they accomplished their assignment explains how grueling this process was.

Pennsylvania's northern border was also under dispute for some time. In the early days of the territory's establishment, that border extended fairly deep into what is now New York because the king granted Pennsylvania's northern border to be all the way to the 43rd parallel. New York, on the other hand, argued the border was supposed to be at the 42nd parallel, and the final decision came down to interpretation of the original charter. The wording was unclear, and each territory pleaded their case.

Again, a compromise settled the matter. Pennsylvania agreed to set its northern border at the 42nd parallel if New York allowed them a notch so they would have access to the Great Lakes. At the northwest corner of the state, surveyors etched out a corner that extended upward, giving Pennsylvania that treasured port city of Erie, their gateway to the Great Lakes.

Chapter 3:
New Jersey

Third state, admitted December 18, 1787

Even though New Jersey officially became a state in 1787, its borders weren't considered finalized until 1833. A 1993 Supreme Court decision gave the boundaries one final tweak.

The original description, back when the Duke of York granted the land to two friends (Sir George Carteret and Lord Berkeley of Stratton), was only loosely defined as "the land between the Hudson River and the Delaware River." The Hudson River turned out to be a bit of a sticky issue. New Jersey claimed their borders should extend to the middle of the river. New York, not feeling particularly generous, argued that New Jersey stopped where the river started.

Some sizable islands were at stake, including Staten Island, which made this a heated issue.

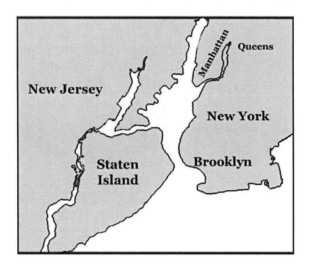

An American revolution came and went, and an arrangement temporarily ended the debate. New York would have ownership of the land east of a line at New Jersey's coast that was *above the water*, and New Jersey would have the land *beneath the water*.

This was such an odd compromise to make and they couldn't possibly have known then that 165 years later, this border was going to be a part of a Supreme Court decision. In 1892, when Ellis Island needed to expand to accommodate the high numbers of immigrants coming into the harbor, New York added to the expanse of the island by dredging soil from underwater. In 1993, New Jersey appealed to the Supreme Court, claiming the newer part of Ellis Island belonged to New Jersey because it had been created from land granted to them in 1833. The land came from *beneath the water*. The Supreme Court agreed, and now that island is split between New York and New Jersey, with New Jersey owning the larger share. (More on that on page 29 in "Water Fights...").

The most unusual feature of New Jersey's borders is the sloping northern boundary. A straight line might have made more sense, but at the time, surveying had a few shortcomings. As part of the early settlement of the region, the Duke of York described the northernmost point of New Jersey as the northern tip of the Delaware River, a body of water that already defined the western border. The technical wording was that the northernmost point of the river was "41° 40' latitude," but the problem is that later surveyors discovered the river actually extends much farther north than previously thought.

When surveyors finally took a serious look at the northern border, they discovered the earlier error in the location of the northwest corner. New York and New Jersey fought over the ill-defined strip of land but finally settled on the deal that dropped New Jersey's northwest border point to its current location. The slope appeared when both parties located the east side border at the point where the Hudson River intersected the 41st parallel.

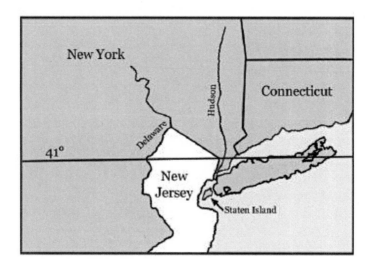

Oddly, some specks of Delaware appear on New Jersey's eastern border. When shipping laborers dredged the Delaware River in the early 20th century to create a lane deep enough for large boats, they dropped the dredged material on the shores of New Jersey. Because that sediment was technically Delaware land, Delaware could claim ownership of those chunks of shore. Neither side has ever made a fuss about it because the land is not habitable.

Chapter 4:
Georgia

Fourth state, admitted January 2, 1788

Georgia is one of several states whose name
originated with the English crown. Honored for
King George II, the colonists constructed the
state of Georgia from land in the charter called the
Carolina Colony. Originally envisioned as an idealized
place for those released from debtors' prisons in
Britain, Georgia's early settlers never saw that vision
realized. Georgia was the last of the original thirteen
colonies and attracted squatters from all backgrounds.

Georgia's borders resulted in bloody battles in both
the north and the south parts of the territory, and one
of the borders continues a dispute to this day because
of water rights. A poor job of surveying resulted in
confusion and debate over exactly where Tennessee
ends and Georgia begins, and the fallout has lasted
hundreds of years. (See page 29 in "Water Fights..." to
find out how poorly surveyed borders can cause

confusion and anger, and become a case for the Supreme Court more than a century later).

When the British first divided the Carolina Colony in 1710, the 35th parallel became the border between North and South Carolina, using the headwaters of the Chattooga River as their guide. As they further broke up the territory to create Georgia, that 35th parallel remained as the preferred guideline for the northern border of the new territory. Unfortunately, surveying in those days was a little shy of scientific perfection, and they incorrectly placed markers 12 miles north of 35 degrees. All sides essentially shrugged at that 12-mile wayward segment because the Cherokee inhabited that land, and white settlers weren't going to live there anyway. But in 1838, a historical travesty that came to be known as the "Trail of Tears" forced the Cherokee off that land. Georgia, the Carolinas and Tennessee all subsequently made claims to the orphan strip.

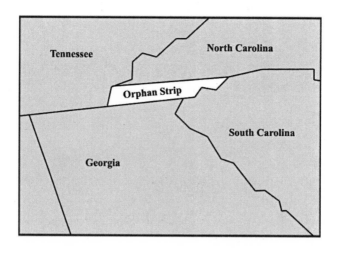

Congress stepped in and settled the matter in 1808. The 35th parallel was indeed the final decree in spite of the earlier surveying snafus. The border remains there to this day.

Creating the western border of Georgia went relatively smoothly, but still came with a quirk. During the colonial period, England sought to make claims to the entire breadth of American land stretching all the way to the Pacific Ocean. Early maps show a rather bold design: the colonies' western borders didn't end until they reached the west coast.

At the end of the Revolutionary War, the territories agreed to break up the long westward stretch of those colonies, and for Georgia, that meant first dividing the section east of the Mississippi River. Territories agreed to donate the excess land back to the federal government, and when Georgia released their additional land, Congress created Alabama and Mississippi.

But that meant Congress needed to decide on Georgia's western border. The Chattahoochee River was an obvious point of division, but halfway up the western side of the state, the Chattahoochee River takes a sharp turn to the northeast. They decided to continue a straight line to the northern border at the 35th parallel. Decisions like that always sound like they make common sense — let's just draw a straight line due north.

But when you look at the map, you'll see that the line doesn't really go straight north at all, but rather angles slightly to the west. In a fortuitous piece of timing, Georgia had just discovered coal in the hills of that region, and by pushing to draw the line at an angle, those rich hills remained in Georgia's borders. This turned out to be a foreshadowing of more state borders as we expanded to the west and discovered gold.

Chapter 5:
Connecticut

Fifth state, admitted January 9, 1788

The name Connecticut comes from a Native American word "Quinatucquet," which means "Beside the Long Tidal River."

Why does Connecticut's northern border have a notch, and why does the southwest corner have a bent leg? And what the heck is "the oblong"?

Much like other states, Connecticut was part of a much larger territory before independence. In this case the larger territory was The Massachusetts Bay Colony. After the Pequot War, which was a bloody confrontation with the indigenous Pequot Indians, settlements spread and the evolution of Connecticut as a separate territory began its course.

In time, colonists in Massachusetts accepted the newly evolved colony of Connecticut, but neither side could agree on its borders.

Massachusetts wanted to locate that border straight west from a point three miles south of Plymouth.

Connecticut argued the line should run straight west from a point three miles south of the lowest point of the Charles River.

The governing body of Massachusetts agreed to the Connecticut principle, but disagreed with which river should be the marker. The resulting settlement drew a line three miles south of the *Neponset River*. But even this trade-off created problems because several well-established towns Massachusetts had claimed fell into this disputed zone, and Massachusetts did not want to give up this valuable land.

Another compromise was in order, and to compensate Massachusetts for the loss of these towns, Connecticut agreed to give up Congamond Lakes, which is why there is a notch in the northern border of Connecticut.

Just to the east of the notch, maps show an oh-so-slight dip where the Connecticut River intersects with the border. This was another concession to Massachusetts, and the dip follows the shape of the hills, thus a nod to the natural geography of the area.

To the west, Connecticut had to work out some kind of arrangement with New York. After much debate and argument, New York insisted its territory extended 20 miles east of the Hudson River, which explains the northernmost stretch of the final boundary, but Connecticut was adamant about keeping its towns of Greenwich and Stamford.

Typical of border disputes, the conflict became ugly, and at one point the governor of New York issued arrest warrants for residents of Greenwich and Stamford. Finally in 1683, they reached a settlement and the pact resulted in a panhandle in the southwest corner of Connecticut where both Greenwich and Stamford are located.

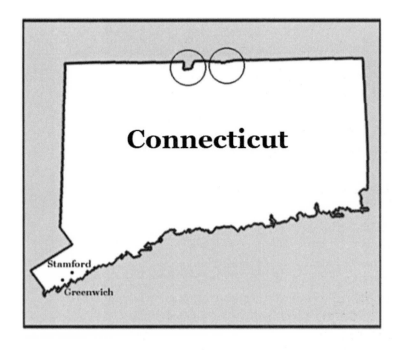

That panhandle, however, required yet another concession to New York because that chunk represented lost land for New York. Connecticut shaved off a long strip of land along the upper western border, turned it over to New York, and this became known as "The Oblong."

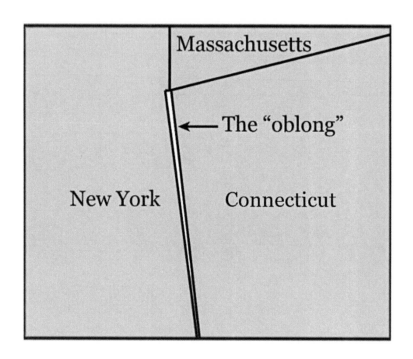

Chapter 6: Massachusetts

Sixth state, admitted February 6, 1788

Massachusetts is one of four states that have retained the title of "Commonwealth," instead of "State." The original name, "State of Massachusetts Bay" remained until the second draft of the state constitution, at which time they adopted the Commonwealth designation. Commonwealth is actually an older term for "Republic" and means the same as "State."

Why does Massachusetts have the land that burps into the Atlantic on its southern border? Why isn't that a part of Rhode Island?

The location of Plymouth is part of what ultimately defined the southern border of Massachusetts. In the original wording of the charter, the southern boundary of Massachusetts was to be three miles south of the southernmost part of Massachusetts Bay, which landed essentially in Plymouth. But Connecticut disputed the claim and insisted that line should actually be eight miles farther north. Townspeople in the area who considered themselves to be under the jurisdiction of Connecticut were not happy with that ruling.

Many years of bickering and wrangling followed and after the Revolutionary War, a commission formed to settle the debate. Massachusetts would include Plymouth and the original settlements — the mass of land that spews into the Atlantic on Massachusetts' southeast border. The outline of their map remained the same even when Congress later carved out Rhode Island. The postwar arbitration then created a line drawn from the southernmost part of the Neponset River, almost straight west.

Two additional concessions appear with a close look at the map. Because Massachusetts would lose several settlements with the new location of that disputed line, Connecticut conceded the area around Congamond Lakes, which resulted in a southbound jog in the border. The agreement also created a jog in the border where the line intersects with the Connecticut River.

(See the map on p. 51 from chapter on Connecticut)

The northern border of Massachusetts went through a series of changes before settling on today's perimeter. Originally set at the 48th parallel, Massachusetts now had to duke it out with New Hampshire, whose charter King Charles II had written. Unfortunately for the two states, the exact border between New Hampshire and Massachusetts was not defined in the charter, so the two territories argued over a three-mile strip for the next 61 years. Finally King George II settled the issue, but his version of the boundary favored New Hampshire, whose residents more closely followed the religious teachings of England at that time.

When surveyors charted out that line, it veered a little to the north as they moved farther west, but in spite of the surveying error, that line stands to this day.

When it came time to divide the land to the west of present-day Massachusetts, New York became the primary debate partner. Massachusetts claimed their western border should be the Hudson River, while New York pointed to the previous, although tenuous, boundary of the Connecticut River as the rightful border. That limit had been good enough for the Dutch, who were the previous residents, so New York felt it should be good enough for the new Massachusetts territory. Again, because this was before the Revolutionary War, England arbitrated the dispute, and mediators set the border between the two states to be a straight line 20 miles east of the Hudson River.

One more quirk in the border needed settling. In the southwest corner of Massachusetts, a small slice of land known as Boston Corners was nestled in the mountains. The only access, because of modes of transport available at that time, was through New York. In a gesture of cooperation, Massachusetts agreed to cede that parcel of land to New York.

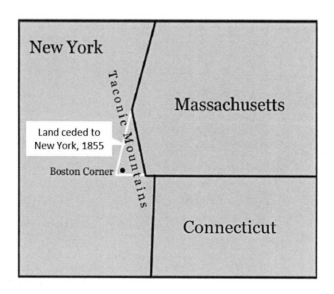

Chapter 7: Maryland

Seventh state, admitted April 28, 1788

M aryland surely must have one of the oddest shapes of all of our states. The ultra-skinny panhandle and the split lower wing with a flared end, can't possibly be the way the state was originally planned. And, in fact, it wasn't. Maryland was originally 1st Lord Baltimore George Calvert's vision as a haven for Catholics. Named in honor of Henrietta Maria, the wife of Charles I of England, Calvert attracted colonists by giving land to the first settlers.

The royal charter described the northern border of Maryland as the 40th parallel, which should have been easy to solidify because Pennsylvania's southern border was also mandated as the 40th parallel. The problem was that the 40th parallel actually cut right through Philadelphia, which was by that time the capital city of Pennsylvania. This discrepancy was one of many examples of how difficult it can be to draw borders sight unseen from across the ocean.

Pennsylvania wasn't giving up Philadelphia, but
Maryland was sticking to the border as dictated by the
royal charter. The dispute took 100 years to resolve,
and led to a border conflict known as Cresap's War.
The intervention of King George II in 1738 finally
relocated the Maryland/Pennsylvania border 15 miles
south of Philadelphia, which explains how Maryland
was stuck with the thin strip we see today.

The dispute with Pennsylvania over its northern
border also created a problem for Maryland on the
east side. The area that is now Delaware was once part
of Maryland Territory, and afforded the state
extensive access to Delaware Bay and the open
avenues for trade that accompanied that access.
Pennsylvania worried its poor relationship with
Maryland would block access to the bay, so they
fought for ownership of Delaware. Lord Baltimore,
representing Maryland, traveled to London numerous
times to argue Maryland's point of view. Finally in
1786, the territories all agreed on an adjustment that
established Delaware and its borders as they exist
today.

To the south, Maryland's border with Virginia
appeared early as the Potomac River (West Virginia
did not exist at that time, and the entire region was
simply "Virginia"). Although the Potomac River
seemed to be an obvious demarcation between states,
problems still arose. Relocating Maryland's northern
border to the line 15 miles south of Philadelphia
(described above) now made it obvious that at one
point the Potomac River flowed so far north it almost
met the new border. This connection nearly cut
Maryland into two parts.

Just west of that narrowing of Maryland, the Potomac splits into a north branch and a south branch. Maryland argued the proper boundary of their state was to follow the south branch of the river, and Virginia argued that Maryland's border should follow the north branch. Once again a land commission needed to resolve the dispute, and Virginia won their side primarily because they were the older of the two colonies. Settlers had already filled the region under Virginia colonization. Because of that decision, Maryland was left with the narrowest of land.

One of the unique features of Maryland's boundaries is the perfect square cut out of the southern border. George Washington selected the land in 1790 to be the center of the United States government in an area now known as Washington, D.C. — the District of Columbia.

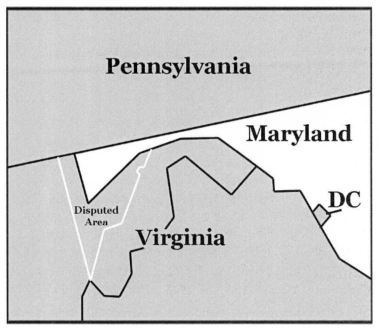

Chapter 8:
South Carolina

Eighth state, admitted May 23, 1788

South Carolina, formed from the territory known as the Province of Carolina, took the name "Carolina," which is Latin for "Charles Land." This territory was among many named for the English monarchy.

From a very early point, the territory of Carolina had two different governing bodies. In the north, new arrivals and transplants from the Virginia Colony had found a home near the Albemarle Sound where farmers grew tobacco and timber. In the south, Charles Town (later changed to Charleston) produced rice and indigo as the prominent crops. The physical distance between the two major settlements, and the difference in cultures between the settlers, made establishing one unified government difficult. In 1710, Queen Anne signed the document that separated the colony into North Carolina and South Carolina, although even with that split, neither territory looked the way it does today because they had originally imagined themselves as colonies that stretched all the way to the Pacific Ocean. Dividing up the extra land would come soon.

Establishing the border to divide the two colonies immediately ran into some snags. The border has a series of angles, jogs, bumps, and curves. At first, cutting the territory in half made sense, which would have meant using the Cape Fear River as a reference. But North Carolina quickly objected to that distinction because they had already issued land titles on both sides of the river, and weren't pleased with the idea of giving up the revenues from those landowners.

The resolution took about 20 years, but the two sides eventually agreed to a new line 30 miles farther south. The intention was to angle the western border up to the 35th parallel, but an error in surveying actually placed the top of that line 13 miles too far south.

To compensate South Carolina for the land lost in that error, England decided that after stretching straight west, the border would jog north when it reached the area populated by the Catawba Indians. Surveyors again made some errors and thus the line extends farther north than the Catawba region. But in the end the change compensated South Carolina for the land they lost in the earlier error.

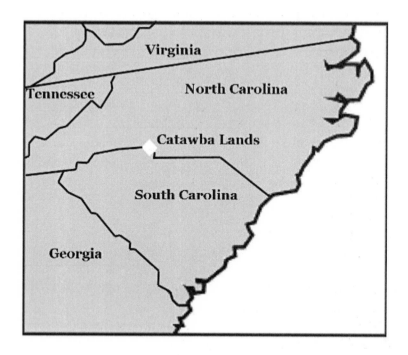

In 1732, King George II established the colony of Georgia, cutting into the endless colonial stretch across the continent, which then solidified the final shape of South Carolina. He used the Savannah River to separate the two colonies, and the final map of South Carolina emerged. South Carolina did not object to having part of its land divided because at the time Spanish Florida was considered a threat to South Carolina's burgeoning plantations, and they welcomed the buffer Georgia provided for them.

Chapter 9:
New Hampshire

Ninth state, admitted June 21, 1788

A s was the case in many of the original colonies, New Hampshire was a territory sliced out of another one. New Hampshire began as part of the Plymouth Colony that later evolved into the territory of Massachusetts. As settlers moved into the area, many accepted the authority of the Massachusetts government, but others began a campaign to make New Hampshire its own colony.

They had timing on their side. The Puritans of Massachusetts were currently embroiled in a head-to-head confrontation with the monarchy of England. King Charles II, intrigued by the notion of diluting Massachusetts' influence by splitting the territory, issued the decree that New Hampshire should stand on its own.

But although he gave New Hampshire its own autonomy, Charles II failed to define its borders. Not surprisingly, New Hampshire, New York, and Massachusetts had different ideas on where to place the new lines, and that debate required almost 100 years to settle.

New Hampshire looked to Massachusetts' royal charter for a starting point. The charter decreed the northeast corner of Massachusetts to be three miles north of the mouth of the Merrimack River. New Hampshire picked that point to be its southeast corner, and then drew a line straight west, all the way to the Hudson River.

Massachusetts wanted New Hampshire to be as small as possible, so their map had New Hampshire extend only to the boundaries of the Merrimack River. When the river took a sharp turn north, so did their version of New Hampshire's border.

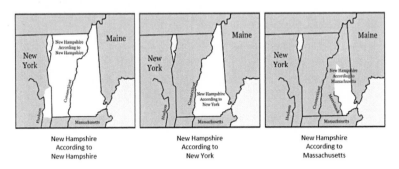

New Hampshire According to New Hampshire

New Hampshire According to New York

New Hampshire According to Massachusetts

King George II stepped in and declared the southern border of New Hampshire should run parallel to the Merrimack River until it turned north, and then continue west as New Hampshire had wished. But allowing the border to continue all the way to the Hudson River cut into land to which New York had already laid claim. Ultimately, in 1763, the king of England (by then King George III sat on the throne) declared the western border of New Hampshire would go only as far as the Connecticut River, a decree that gave New Hampshire fits because their residents had already settled west of that river.

England threaded a fine line between the many battles involving its American Colonies. They had also recently acquired Quebec, so they felt compelled to make a strong statement with regard to New Hampshire's desire to expand. If they could stem that ambition, the king wouldn't have to worry about a conflict with Quebec and the important water routes that natural resource afforded England. Much later, this decision played a part in forming Vermont.

In the meantime, the king dealt with the delineation of New Hampshire's eastern border. Because early land developer Captain John Mason had made claim to the stretch of the Piscataqua River that meets up with the Atlantic Ocean, this spot became the starting point for the eastern border. The agreement ultimately made the border follow the Piscataqua and Salmon Rivers north until reaching the headwaters of the Salmon, and then continue due north for a total of 120 miles.

Chapter 10: Virginia

Tenth state, admitted June 25, 1788

Virginia was the first established colony, but didn't become a state until nine others had joined the Union. Early settlements in Virginia met with great adversity, and some completely disappeared as settlers had poor provisions. The introduction of tobacco as a crop, and slaves as labor, changed all of that, and the new colony began to prosper.

At the time of colonization, many borders still had rather loosely defined descriptions, and Virginia's perimeter was a good example. The chunk of land stretched between 34° and 41° latitude, and extended all the way from the Atlantic Ocean to the Pacific Ocean. Modern-day Virginia looks considerably different.

When Maryland entered statehood earlier in 1788, Congress defined its southern border as the Potomac River. As a result, Virginia's northern border was essentially set, but the two territories argued over which branch of the river would be used. The Potomac split into north and south branches, a distinction that created vastly different notions of who got which piece of land. Virginia's interests prevailed because they had already deeded some of the disputed land to Virginia colonists.

Among the many goofy little quirks of our states shapes is a parcel of a peninsula at the bottom of Maryland's territory that actually belongs to Virginia. The property is like a finger of land that hangs off of Maryland, and whose residents peer across the bay at their home state of Virginia. In those early days, without solid borders to give guidance, territories often issued deeds to property that wasn't actually theirs to sell. Colonists from Virginia ventured across Chesapeake Bay and began working the land at the tip of the peninsula that was actually part of the Maryland territory. Maryland argued the entire peninsula was part of their claim, but colonists had *already paid* their money to Virginia for the rights to the land. As had happened in many other land disputes, the English government stepped in to mediate. The final decision was weighted toward the colony that had sold the land first, so Virginia won the dispute by default.

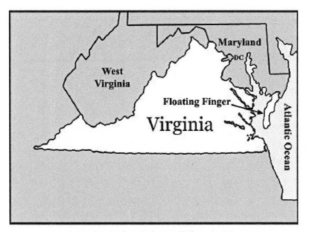

Virginia's Floating Finger

If you've been reading this book chronologically, you will have noticed that Maryland lost every one of their disputes with other states.

Surveying errors explain the slight rise to Virginia's southern border. Debates with the Carolina Colony eventually resulted in a description of that southern border as running straight east/west slightly north of the 36th parallel. Virginia and Carolina argued about who had tariff rights in the Albemarle Sound, and King Charles took Carolina's side on that issue. This decision placed Virginia's southern border one-half a degree north of the 36th parallel, but the line actually curves slightly up as it moves west. The result is that at its western end the border rises five miles north of the 36th parallel. The surveyor of the errant western end of that line was Thomas Jefferson's father Peter Jefferson. Some of the surveying errors at the time were vigorously challenged. This one was allowed to stand.

Virginia's western border wasn't delineated until Congress created West Virginia and Kentucky. Virginia responded to a request from the new United States government for colonies with large landmasses to cede back portions to the federal government. The border defined as the separation between Virginia and the new Kentucky territory was a meandering line that followed the Appalachian Mountains and then turned sharply when it reached the Tug Fork River. There the border stretched until it intersected the Ohio River.

Virginia's western border was to change one more time. At this point Virginia's territory still included the land that later became West Virginia (see more in the chapter on West Virginia). Once Virginia and West Virginia successfully separated, Virginia's western border was set.

Virginia Before the Split

Chapter 11:
New York

Eleventh state, admitted July 26, 1788

When the Dutch came to the New World in the 1600s, they claimed a large portion of land they called New Netherland. Seventy years later the British defeated the Dutch and began the process of creating the state of New York from the territory they won. King Charles II gave the land to his brother, the Duke of York, and the name was born.

As with most early border decisions, vigorous debates preceded the decisions about final boundaries. New York argued their boundaries should extend as far east as the Connecticut River because that had been the landmark the Dutch used. Connecticut argued vehemently against that because such an extension of New York would have stolen their established towns of Greenwich and Stamford. Each territory fought for port access to Long Island Sound, which meant any acceptable compromise would have to include access for both.

By locating the eastern border of New York ten miles east of the Hudson River, both sides could be soothed to a certain extent, although they ultimately made a further concession to create the small panhandle in the southwest corner of Connecticut to preserve Connecticut's existing settlements. To compensate New York for that accommodation, the line that rose up between the states included an extra sliver, cleverly called "The Oblong," in New York's favor.

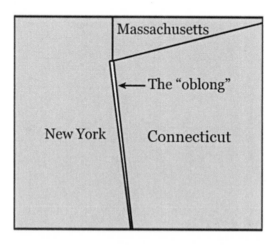

North of Connecticut, New York still had to deal with Massachusetts. New York insisted its border should extend to the Connecticut River, but Massachusetts made the claim that New York could only spread as far east as the Hudson River.

The British monarchy finally stepped in and drew the line that exists today, essentially continuing the border established between New York and Connecticut. The exception is the little notch taken out of the southwest corner of Massachusetts. This was a hilly, almost unreachable terrain that couldn't be accessed or policed from the Massachusetts' side of the area known as Boston Corners. Happy to rid itself of the rather unruly township, Massachusetts redrew the line so that the notch became part of New York.

New York once again had to go to the compromise table over its border with New Jersey. For a period of time, both territories made claim to a strip of land between the 41st latitude and the Delaware River, and predictably those who settled in the region were less than thrilled two different territories charged them for deeds and billed them for taxes. Eventually the two sides reached an agreement by drawing a line from where the Delaware River takes a 90° turn to the north, on a straight line angled to where the Hudson River meets the 41st latitude. At that point, the river and the bay were split, and the territorial government creatively drew a line around Staten Island. Their decision turned out to have ramifications more than 100 years later, as explained further in the chapter on New Jersey.

New York and Pennsylvania also argued for quite some time about the location of their shared border.

Eventually Pennsylvania became willing to find an accommodation with New York when, to the south, Virginia conceded some land to Pennsylvania. In the spirit of compromise, Pennsylvania agreed to New York's claim the border should fall on the 42° line, with a nifty notch cut out where the line meets Lake Erie so Pennsylvania could enjoy a larger port exposure.

Much earlier, in 1763, New York had agreed to its border with Canada. The St. Lawrence River had already become an important means of transporting large amounts of commerce, and neither side would concede the waterway to the other. The St. Lawrence River became the natural boundary between Quebec and New York, giving both sides complete access. They also agreed where the new boundary crossed the 45th parallel, the line would turn straight east to the Connecticut River.

Chapter 12:
North Carolina

Twelfth state, admitted November 21, 1789

A s discussed in the chapter on South Carolina, the larger territory of Carolina began to split ideologically into two separate entities as early as the late 1600s. The actual legal process to separate North Carolina and South Carolina began in 1710.

The original Carolina charter located the northern border of the new territory at the 36th parallel, but settlers immediately sent word to the king of England that an adjustment had to be made. The 36th parallel ran right through the middle of Albemarle Sound, which was the primary settlement spot. Splitting the Sound caused settlers great financial distress because Virginia charged them shipping tariffs for all the commerce they moved through the waterway.

Luckily for North Carolina, the king responded to their pleas by moving the boundary a little farther north, which is why the current and permanent northern border of North Carolina is halfway between Albemarle Sound and Chesapeake Bay. With this deal, both territories had plentiful access to commercial waters.

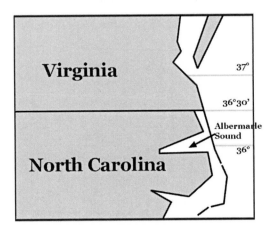

An early consideration for dividing the two Carolinas included using the point at which the Cape Fear River met the Atlantic Ocean, but by the time discussion had gotten this far, North Carolina had already been issuing deeds to land on both the north and south sides of the river. They crafted a new decision to locate that border point 30 miles farther south, and then run that new point northwest until it reached the 35th parallel. Under the original wording of the contract, the border was then to turn and head straight west.

A border like that should have been simple, but as the early surveyors and legislators found, nothing ever quite turns out as planned.

The British hierarchy was very aware that part of the area their lines would cross would be land belonging to the Catawba Indians, and farther west, the Cherokee. Early settlers already had a great deal of conflict with natives as the newcomers made claims to land. With the notion of staving off unnecessary bloodshed, surveyors made adjustments when they encountered Catawba land. The idea was to draw lines around indigenous claims and therefore avoid conflict.

But an earlier surveying error made that a moot point. The line that was supposed to run northwest to reach the 35th parallel had unintentionally stopped 13 miles south of that marker. When surveyors continued to the west, they never ran into Catawba Indians or their land.

While they avoided one problem, they created another. By surveying too far south, the surveyors cheated South Carolina out of some of the land that was supposed to be their territory. The British government made yet another adjustment, and thus the border lost its intended straight western line and instead veered north and followed the Catawba River until it branched. At this point, the line continued due west, although again the surveyors and their equipment were less-than-perfect and the line actually veers a bit north before the correction is made to bring it back to the 35th parallel.

When the monarchy first imagined the Carolina territory, the western border was designed to stretch all the way to the Mississippi River. But after the American Revolution, the federal government made a plea to these long-stretching territories to donate land west of the Appalachian Mountains so they could produce new states from that land.

In the case of North Carolina, that lopped-off piece of land became Tennessee (and for a brief time, also the State of Franklin. See "States that Didn't Make the Final Cut" on page 23 for more information). For the most part, the crest of the Appalachian Mountains formed that new border.

Chapter 13:
Rhode Island

Thirteenth state, admitted May 29, 1790

At first thought it appears odd that a state that is not actually an island has the word "island" in its name. Even odder is that the true name of Rhode Island is: *The State of Rhode Island and Providence Plantations*. The smallest state has the longest name.

The explanation for both comes from the fact the state we know today resulted from a merger of two colonies. Providence Plantations was a colony in the area of the modern city of Providence, established in 1636 as a site of religious freedom. Rhode Island Colony was founded two years later on the largest island in the adjacent Narragansett Bay. At that time, the island's name was Aquidneck, but was later changed to Rhode Island.

By 1647, the two large settlements, along with two other small ones, agreed to band together in order to protect themselves against other colonies. They united under "Charter of Rhode Island and Providence Plantations," the predecessor of the state name.

Rhode Island was in large part formed from Connecticut territory, but not, as you can imagine, with Connecticut's blessing.

When settlers originally established Connecticut, the eastern border extended all the way to Narragansett Bay. In order to create Rhode Island, the British pushed that border all the way back to where the Pawcatuck River met the Atlantic Ocean. Connecticut, of course, fought this big bite taken out of its own charter, but by 1840 they relented and accepted the new territory.

Commissioners designated Rhode Island's northern and eastern boundaries as the southern, angled sides of the Massachusetts territory. Connecticut and Massachusetts had already bickered over exactly where these lines should fall. King Charles II simply transferred all that hard work over to Rhode Island's new charter, thus giving the state its current shape.

But as was typical in those days, lines were still a bit ill defined. Because surveying was a slow process back when Connecticut still owned that land, Massachusetts had already issued deeds to land and made claims south of the presumed border. Rhode Island inherited some of those land disputes, and ultimately agreed to define its northern border slightly south of the charter directives, and then angled northward as it progressed east.

This is why the Connecticut and Rhode Island borders do not match up exactly at Massachusetts' southern edge.

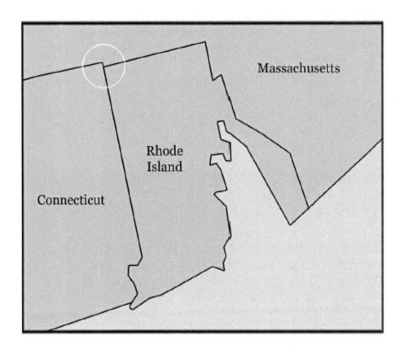

Rhode Island still had some things to work out with Massachusetts. On its eastern side, they had a "big picture" outline, but needed to agree on exactly where Rhode Island stopped and Massachusetts began in Narragansett Bay. This time King George II stepped in to make the declarations, but his choices didn't sit well with either side. In the end they went through a series of land swaps, and finally stamped their pact with the boundaries that remain to this day.

As for the name of Rhode Island? As recently as 2012, voters had an opportunity to weigh in on the issue of whether to keep the historic, but lengthy full name of *The State of Rhode Island and Providence Plantations* or to fall in step with the rest of the country and shorten the name to the simpler "Rhode Island." By an overwhelming vote, citizens chose to keep their longer, more eloquent, historical name.

Chapter 14:
Vermont

Fourteenth state, admitted March 4, 1791

The existence of Vermont is owed in large part to the militia group called The Green Mountain Boys, led by Ethan Allen. Snuggled neatly between New Hampshire and New York, both states claimed ownership of the territory, then known as New Hampshire Grants. When New York sold leases to land already settled by people who had bought grants from New Hampshire, battles to defend their land followed. The Green Mountain Boys were born as a way of defending the settlers. After years of resisting New York's claims to land they had settled, they created the new territory of Vermont in that space. Although they disbanded a year before Vermont was granted independence as a territory, the militia group has been widely credited with the existence of Vermont as a separate state.

The dispute certainly caught the attention of George Washington, who considered sending troops to overthrow the government that Vermont ultimately installed. In the end, all sides calmed down and Vermont kept its status as a Republic until joining the Union in 1791.

Agreeing to new boundaries is tough when all sides are at odds with each other, but after years of negotiation, arrangements were ironed out. The establishment of Vermont occurred at roughly the same time the British were eager to solidify the line between the U.S. and Canada, and England had incentive to create a peaceful demarcation between two groups of people who had no interest in getting along. French Catholics previously settled Quebec, and British Protestants found a home in New York and nearby territories. Both sides had an interest in a hard border that would be sure to preserve access to the St. Lawrence River, a major commercial "highway" at the time.

With that in mind, the 45th parallel became the northern boundary of New York and, by default, Vermont's northern border too.

To determine Vermont's western border, commissioners made the decision to continue the 20-mile-wide buffer east of the Hudson River that satisfied the two sides in the New York/Massachusetts dispute (see the chapter on Massachusetts). As it extended farther north, the edges of Lake Champlain sufficed to define Vermont's western border.

On the east, the Connecticut River became a natural demarcation between Vermont and New Hampshire.

Much earlier, New York tried to claim all the land west of the Connecticut River. In a sense, Vermont just slid into the space New York had already earmarked as theirs. As mentioned, none of the sides happily agreed to the carving up of the land, but through trade-offs and a little strong-arming, both sides made tough decisions and Vermont gained statehood.

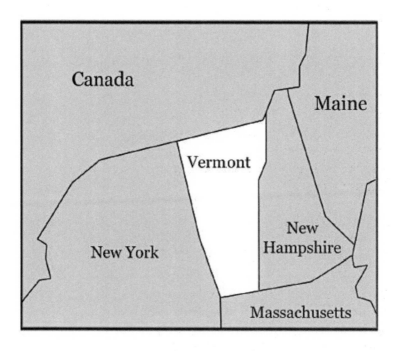

Chapter 15: Kentucky

Fifteenth state, admitted June 1, 1792

I n the 1700s, people traveled primarily by horseback or wagon, or if you were lucky enough to be near water, by boat. Mountains were a massive impediment to travel and created a natural separation between peoples.

The federal government challenged Virginia to spinoff part of its western territory, and when they did, the Appalachian Mountains created a natural boundary. Rivers were also a common natural boundary so for Kentucky, mountains and rivers account for most of its shape.

The Cumberland Gap is known in historical lore as the gateway through which Daniel Boone led many groups of settlers and explorers. Accordingly, the Cumberland Gap provided a natural starting point to carve out the new state of Kentucky from Virginia. The line basically followed the crest of the mountains as it made its way north. Where the border met with the Tug Fork River, it turned and followed the river north.

The southern borders of the future states of Ohio and Indiana had already been fixed because they were part of the Northwest Territory, defined as the land north and west of the Ohio River. As a result, the Ohio River became the meandering boundary that we now recognize as Kentucky's northern boundary.

The tiny piece that is Kentucky's western border is defined by the Mississippi River. At that time, France still controlled the region west of the Mississippi, so American settlers had no plan to extend claims beyond the mighty Mississippi.

The southern border of Kentucky contains a couple of oddities. The original charter located that demarcation as a straight line at 36° 30', an expression of location that is read "36 and a half degrees north of the equator," or, alternately, "36 degrees and 30 minutes north of the equator." This locator should have been simple and straight and uninteresting because it was a continuation of Virginia's southern border. But nothing was simple or uninteresting in those times.

While Virginia territory hacked off parts of its land to create Kentucky, North Carolina was doing the same to parts of its land to form Tennessee. Virginia hired a surveyor named Dr. Thomas Walker to survey the southern border of Virginia's new offspring (Kentucky), but his line unintentionally veered to the north rather than follow the instructions in the charter. This was just one in a long line of surveying errors that gave territorial governors ulcers.

One hundred years of wrangling between Kentucky and Tennessee followed, and the time was spent swapping numerous deals. Precision in surveying continued to be a problem, however, and the final agreed-upon line still didn't match exactly where Virginia's southern border ended, thus creating a very small discrepancy in Virginia and Kentucky's southern borders. And it still angled slightly to the north. Eager to finally settle their disputes, both sides accepted the flawed line.

In 1819, General Andrew Jackson bought some land from the Chickasaw Indians and divided it between Kentucky and Tennessee. In that agreement, Tennessee's portion of the land went only as far north as the 36° 30' line as originally charted for the entire border. The result was that Kentucky's piece of that land drops down at their southwest corner. The dangling chunk gives the appearance of something added as an afterthought, which is essentially what it was.

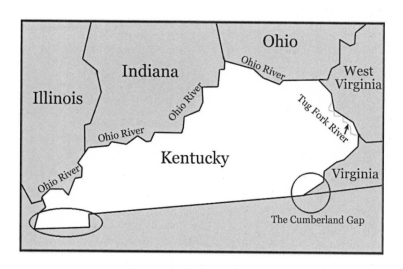

Chapter 16:
Tennessee

Sixteenth state, admitted June 1, 1796

At different points in Tennessee's history, the Cherokee Indians, Chickamauga Indians, Chickasaw Indians, France, Spain and Britain all laid claim to the land. Not until the Treaty of Paris in 1783 did the British officially declare proprietorship and begin to establish boundaries.

Tennessee was originally part of the larger Carolina Colony, but in the late 1700s, the federal government asked some of the larger colonies to give up parts of their land to create new territories. Because governing land on the other side of the Appalachian Mountains was very difficult and expensive, it made sense for Carolina to use that natural feature as the boundary between the old and the new territories.

Initially, several counties in the northeast part of what is now Tennessee declared themselves to be an independent state they named Franklin. Citizens elected a governor, adopted a constitution, and conducted business as a sovereign state, but financial difficulties resulted in North Carolina reclaiming Franklin after four years. In the end, the future state of Tennessee absorbed Franklin. See "States that Didn't Make the Final Cut" on p. 23 for more on the State of Franklin.

The highest crests of the Appalachian Mountains defined most of Tennessee's eastern border, with the curious exception of the southeast corner where the line drops straight south. Governments of all of the new territories used land sales as a means of raising funds to manage the many obligations to its citizens, but often those territorial boundaries were not well defined. Thus, land grants were commonly issued from government entities on land that was later found to belong to another territory. This happened in the mountainous areas between North Carolina and Tennessee, and, although the details are unclear, the straight line dropped south appears to have been a compromise worked out to avoid a long and expensive fight.

Tennessee's southern perimeter, the 35th parallel, is simply a continuation of the border previously designated for North Carolina, although they never did get it quite right. See the chapter on "Water Fights," p. 29, to see why this is still an issue in the 21st century.

The northern border of Tennessee caused the most anguish between the two sides and the surveyors. Although the charter had proclaimed 36° 30' to be the line to separate Tennessee from Kentucky, events of the time got in the way of successfully marking that border. Surveying errors resulted in a line that gradually angled north as it moved west. By the time the line reached the Tennessee River, they were considerably off the mark.

When Andrew Jackson bought land from the Chickasaw Indians just west of this point, this surveying error once again came to the forefront. He intended the newly claimed land to be split between Kentucky and Tennessee, but with a poorly executed surveying job, the actual boundaries lay in dispute. After much negotiation, Tennessee accepted the new piece of land and, along with it, the anomaly in the northern border that gave that little drop-down piece to Kentucky.

Chapter 17: Ohio

Seventeenth state, admitted March 1, 1803

The Ohio Territory was originally part of a larger landmass called The Northwest Territory, so-called because of its location north and west of the Ohio River. That landmass would later become Wisconsin, Michigan, Indiana, Illinois and Ohio. Statehood was granted to each territory as they reached a population of 60,000.

Determining Ohio's western border was the easy part. Beginning at the point where the Ohio and Great Miami Rivers met, commissioners drew the line straight north. Where the line stopped to determine the northern border became a bit more prickly.

In the wording of the Northwest Ordinance, which spelled out the rudimentary division between the territories in The Northwest Territory, Ohio and

Michigan were to be divided by a line that extended straight east from where the Maumee River meets Lake Erie. This is where the problem arose. That meeting point happened to be right in the middle of Toledo, which had already been established as a burgeoning port city and was crucial to the commerce of the Ohio Territory.

Michigan also wanted that land and port. When Congress redrew the line slightly north of Toledo to favor Ohio, the Michigan representatives spoke up. In fact, they spoke up with guns in hand and prepared to shed blood. The Toledo War, as the pending skirmish came to be known, never really amounted to much because Congress stepped in again and forced a settlement. This settlement conceded Toledo to Ohio and, in compensation, granted Michigan the Upper Peninsula in the north. This is the answer to the question: "How did Michigan end up with the Upper Peninsula?"

East of Ohio, Pennsylvania had already been admitted into the Union and had a nice straight western border. One would expect Ohio's eastern border to meet Pennsylvania's border so the two states would share a straight line, but a look at the map shows that the straight border only goes part of the way for Ohio.

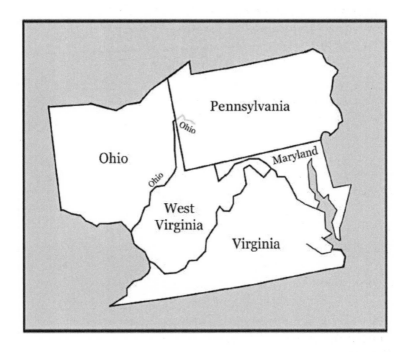

Virginia owned the sliver of land—which would later become part of West Virginia—that filled in between Pennsylvania and the Ohio River, so the division between Ohio and Pennsylvania would hit a bump. Ohio's eastern border *begins* as a straight line shared with Pennsylvania, but deviates farther south to follow the Ohio River, and the Ohio River also defines the southern border. Interestingly, even though the river defines much of Ohio's border, the river itself is part of Kentucky and West Virginia. This had to do with the original wording that described the border as the northern low-water mark, meaning the vast majority of the river itself falls within the boundaries of neighboring states.

Chapter 18:
Louisiana

Eighteenth state, admitted April 30, 1812

L ouisiana has an odd shape and a unique, amalgamated history. What was first called "Louisiana"—named after French King Louis XIV—was the massive landmass we know today as "The Louisiana Purchase." Included in this mass were the states that later came to be known as North Dakota, South Dakota, Minnesota, Wisconsin, Michigan, Indiana, Illinois, Iowa, Nebraska, Kansas, Missouri, Oklahoma, Arkansas, Mississippi and Louisiana.

So how did this giant territory become an average-sized state shaped like a boot?

Thomas Jefferson, from the very beginning, pushed the concept of dividing states into smaller, but fairly equally portioned entities. He believed if individual states were too large, they would collapse under the weight of the difficulty of governing a large landmass.

He also believed if states were relatively equal in size they would enjoy equal representation in Washington, D.C. and have the potential for similar amounts of natural resources. See the chapter called "Jefferson's Ideal of Equal States," p. 31 for more on this.

Jefferson also saw a value in keeping a state small enough to maintain a certain homogeneity. The French settled much of the area that became Louisiana, and when the time came to define its boundaries, keeping those French settlements within the same state was an important factor.

One of the reasons the expanding United States was interested in the larger Louisiana Purchase territory was the enormous value of the throughway known as the Mississippi River. As a country, access to the river was imperative, so the river became a natural focal point for determining boundaries when they divided the territory into smaller portions.

The budding state of Louisiana had Spanish-held land on all sides, which influenced its shape. At that time, Florida was a Spanish territory, and its panhandle stretched all the way west to the Mississippi River. Spain also owned much of the land west of the Mississippi River Basin, so carving out Louisiana meant negotiating and even fighting with Spain. Not until the U.S. seized the land between the Mississippi River to the Pearl River from Spain did Louisiana acquire the part of its land shaped like the toe of a boot.

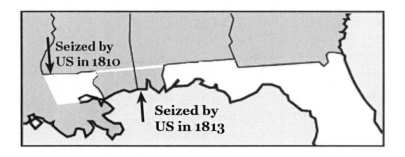

Negotiating Louisiana's western border with Spain didn't require as much military might. Spain was embroiled in colonial uprisings in both Central and South America at the time so war weariness in their ranks played a part in Spain's decision to settle this dispute rather quickly.

By 1819, the U.S. and Spain reached an agreement that finalized the Louisiana boundary we see today. The toe of the state turned out to represent the first stage in reducing Spain's control of the Florida panhandle to modern-day dimensions.

Chapter 19:
Indiana

Nineteenth state, admitted December 11, 1816

I ndiana was part of a larger landmass known as
The Northwest Territory. This territory had been
the nation's first attempt at expansion to the west,
and the federal government wanted to make sure
divisions of the land would be equitable and peaceful.
They wrote the Northwest Ordinance to describe the
criteria for creating new states within The Northwest
Territory.

Northwest Territory

One of the criteria was that a territory had to achieve a population of 60,000 in order to become a state. Indiana began their road to statehood in 1800.

The Ohio River defined the southern border of Indiana back when Congress first created The Northwest Territory. In fact, the original name of the territory was "The Territory Northwest of the River Ohio" and was an early American effort to push back on an English ruling designed to discourage settlements west of the early colonies. England had been worried that if America got too big, she would be difficult to control.

Ohio became a state before Indiana, and therefore the straight north/south line from the place where the Great Miami River becomes a tributary of the Ohio River was already established as Indiana's eastern border. But how far north was Indiana's border supposed to go? This turned out to be a bit of a problem.

Originally the northern border of Indiana was simply a straight line running east to west that would meet up with the very southern tip of Lake Michigan. But Indiana was adamant they needed more access to the shipping opportunities of the Great Lakes, and they lobbied Congress for a border a little farther north. Just a tiny tip of a great lake was not sufficient, but moving that border meant grabbing up some of the land the territory of Michigan laid claim to. Naturally Michigan mounted a strong opposition. The territory known as Michigan had already conceded a similar stretch of land to Ohio, and they weren't in the mood to give up more.

But in the end, Michigan only had the rights of a territory. They would not become a state for another 21 years and therefore had no real vote in the matter. Indiana's claim won, and Congress repositioned their border ten miles north, giving Indiana port access to the lake.

Congress was already looking ahead to the formation of Illinois when they negotiated Indiana's western border. Eager to make both states about the same size, they decided the Wabash River made a good starting point. At the point where the river began to veer more to the east, they drew the line straight north until it reached Indiana's northern border.

Forming Indiana's borders may possibly have been one of the most straightforward endeavors in U.S. history.

Chapter 20: Mississippi

Twentieth state, admitted December 10, 1817

Although located in the general vicinity, Mississippi wasn't part of the giant Louisiana Territory purchased from France in 1803. French explorers were very active in the area and traveled the entire stretch of the Mississippi River early on, establishing French settlements along the way.

The territory of Mississippi had once been part of the Carolina Territory. As was the accepted procedure at the time, huge territories like Carolina agreed to give up parts of their holdings and divide the land into mostly equal-sized states. Because the early colonists considered their land rights to extend all the way to the Pacific Ocean, the Carolina Colony contained all or parts of what later became North Carolina, South Carolina, Georgia, Tennessee, Alabama, Mississippi, Arkansas, Louisiana, Oklahoma, Texas, New Mexico,

Arizona, and Southern California. Breaking it up made sense.

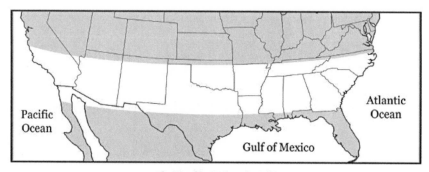

The Carolina Colony in 1629

Mississippi's shape changed several times before statehood. The first version of the state contained the land that later encompassed both modern-day Mississippi and Alabama, and its southern border stopped at the 31st parallel. At that time, Spain owned the long panhandle-shaped territory below the 31st parallel, so territories like Mississippi and Alabama had no direct access to the gulf. In 1813, the U.S. Congress seized the panhandle land east of the Pearl River, which explains why Mississippi has a tab shape on the southern border butting up against Louisiana. Louisiana had already laid claim to the rest of that piece of panhandle west to the Mississippi River.

In the early days of the creation of territories and states, France still had fur trappers working out of the northern part of modern-day Mississippi, so the original Mississippi territory did not reach as far north as its current boundaries. Once the French left the area after the Louisiana Purchase went through, Congress felt free to extend Mississippi's border all the way north to the 35th parallel, which corresponded to the northern border of Georgia. All three modern states, Georgia, Alabama and Mississippi, share the 35th parallel as their northern border.

The next decision was to divide the large territory essentially in half and create Alabama. But the line dividing the two states isn't straight. Was this intentional, or did the mapmakers have trouble drawing a straight line?

As it turned out, that bend in the dividing border was intentional. Congress used the area where several rivers flowed directly into the Gulf of Mexico as a marker to create the division for the southern half of the combined Mississippi/Alabama territory. The soil around the rivers was excellent for farming, and the fair decision was to divide this land up evenly between the two domains.

But by following this line, Congress slightly favored Alabama in the acreage count, so as the dividing line continued north, they intentionally curved it back slightly to the east to give the two territories a fair distribution of land.

Why did the Congress decide to take a little snip from the northeast corner of Mississippi and grant it to Alabama? This is the spot where the Tennessee River angles near the border. Had they continued the dividing line straight north that decision would have created a tiny area of Mississippi cut off by the river and separated from the rest of the state—so the border was rerouted. People who had settled in that small area became residents of Alabama.

Chapter 21: Illinois

Twenty-first state, admitted December 3, 1818

T he borders of Illinois have been altered on a number of occasions, sometimes by the power of politics, and at least once by the power of nature.

When Illinois became a territory, they established the capital at Kaskaskia. You've never heard of Kaskaskia, Illinois? Perhaps because Kaskaskia hasn't been the capital of Illinois since 1819, it currently has only about 14 residents, and many years ago the town was literally detached from Illinois.

Set on the banks of the Mississippi River, Kaskaskia became the victim of the power of the changing course of a mighty river. The need for wood for building caused massive deforestation of that stretch of the riverbanks and, in 1881, floods destroyed most of the town. When the waters receded, the river had redirected itself into a new channel, functionally separating the town from the state of Illinois. On the western side of the town where the Mississippi River used to run, now lay a dry riverbed and the state of Missouri. Almost 200 years later, the river remains diverted to the east side of the town.

The United States Postal Service assigned Kaskaskia a Missouri zip code because carriers were only able to access the small town from that side. A few of the town's original buildings stand, and most of the land is now dedicated to raising crops.

By the time Illinois entered the Union, its eastern border were already determined by Indiana, ratified as a state two years earlier. The western border was the Mississippi River and the southern border the Ohio River, both delineations spelled out in the Northwest Ordinance that described the boundaries of The Northwest Territory. Illinois' northern border would go through two distinct adjustments.

Because The Northwest Territory also included Michigan, Wisconsin, Indiana and Ohio, the original northern border of Ohio, Indiana and Illinois was a straight line. This decision was part of Jefferson's ideal that all states should be approximately the same size, yet at different times, both Indiana and Ohio negotiated changes in their northern borders to afford them better access to the Great Lakes. Tweaks in their borders also resulted in changes for Illinois.

Illinois wanted the same thing that had motivated Indiana and Ohio's border disputes. They argued for a greater exposure to Lake Michigan to make commerce and transportation more viable. They were particularly interested in devising a canal system that allowed them to move goods from the Mississippi River to Lake Michigan to help them become a center of commerce. In order to gain the amount of land adequate for these canals, Illinois convinced Congress to locate their northern border 60 miles farther north than the original plans. Illinois opened the canals in 1848, and the redrawn border is what you see on modern-day maps.

Chapter 22:
Alabama

Twenty-second state, admitted December 14, 1819

Alabama looks almost exactly like Mississippi but in a mirror image. Previous to Alabama becoming a state, the country had admitted 21 states to the Union and not one of them looked like any of the others. Why did Alabama end up as a carbon copy of another state?

Alabama and Mississippi were both chiseled out of the larger landmass known as the Georgia Territory at a time when the government believed we would be better served having a greater number of small states, rather than a handful of very large ones. Massive territories were cut into smaller ones, and thus the "Mississippi Territory" was born. Sometime later Alabama was carved from this new territory.

The government needed to establish the exact location of the border between Alabama and Georgia, and in typical fashion of the time, a major river became the focal point. The Chattahoochee was a good starting point, beginning at the point where United States land met Spanish land on the long Florida panhandle. But eventually the Chattahoochee takes a turn to the east, so at that point Congress made the decision to change the line to one that ran straight north to meet the 35th parallel. The 35th parallel was the northern border that had already been established for the larger Georgia territory.

Yet as you'll notice, that straight line actually veers to the west, not true north. Georgia laid claim to the rather abundant coalmines in that region of the Appalachians, and insisted the borderline that created Alabama had better steer clear of those mines, thus the bend to the west.

But why create a little notch for both Alabama and Mississippi on their southern ends?

For many years, Spain's claim to Florida included a long panhandle that extended much farther west than the panhandle of Florida we know today. Alabama's southern border would have been a simple straight line except for the fact that when the War of 1812 was approaching, Spain lost the ability to focus on their ownership of Florida, and relinquished its hold on parts of the panhandle, although not entirely willingly.

Congress broke up the tracts of land in the panhandle it had seized and awarded one of those chunks to the Mississippi Territory. When they divided that land with a line down the middle to become the states of Mississippi and Alabama, the notches in the southern boundaries of both states appeared.

Chapter 23: Maine

Twenty-third state, admitted March 15, 1820

When you look at the map of the United States, Maine looks like a late add-on, as if the original settlers wanted to tack on another state to extend as far into the other British territory (Canada) as they could. Maine wasn't one of the original thirteen colonies, so how did it appear where it did?

The state of Maine originally did begin in 1620 as part of the Plymouth Colony, later renamed the Massachusetts Colony. Much later, while settling the Maine border, plenty of people had differing opinions on where it should stand. At stake to the north was the very definition of what constituted the United States, and American interests were busy creating settlements as far north as they could. By pushing north, they deprived the British of as much Canadian land as they could swipe.

For the British establishing territories in Canada, the St. Lawrence River was a jewel in their crown. They had no intention of allowing the despised revolutionaries to cut them off from this important lane of commerce. In the end, they were successful in that goal.

The early versions of the treaty contained language that was vague and unclear, so the task of settling the border dispute was ultimately turned over to the Dutch. But neither side particularly liked the arrangement the king of the Netherlands proposed, and they continued to bicker, mostly because both sides coveted the rich timberlands.

Before they reached a final settlement, a border skirmish called the Aroostook War (named after the valley) broke out. The war was bloodless, although there are rumors the conflict did bring one casualty: a pig that wandered across contentious lines. In 1842, the famous Daniel Webster finally brokered a treaty to define the northern border we recognize today. Americans probably should have taken the deal offered by the Dutch king because that version would have given the U.S. a bigger chunk of land than the ultimate Webster treaty, but appeasement came in the form of receiving access to the northern reaches of the St. John River.

The western border of Maine had been determined when New Hampshire became a state. As was the custom of the time, the boundary began with a river, in this case the Puscataqua, and when the river reached its headwaters, the line continued straight north. Observers will note that it actually veers two degrees to compensate for the curvature of the earth.

Chapter 24: Missouri

Twenty-fourth state, admitted August 10, 1821

In 1818, a man named John Hardeman Walker jumped on his horse and began the long journey from the territory of Arkansas to Washington, D.C. He intended to strong-arm members of Congress into making a decision that would greatly benefit him financially. By all accounts his efforts paid off. Today we call what he did "lobbying." Walker called it good business.

As both the Arkansas and Missouri territories approached statehood, the line dividing them was a straight one. No bootheel existed or was even imagined, but cattle played a part in how that was formed.

Walker owned a cattle ranch and had grown his property to an enormous size that comprised a large part of the northeast corner of the Arkansas Territory. He watched with some anxiety as Missouri got a head start on establishing towns and expanding wealth. St.

Louis was already a thriving gateway destination, claiming thousands of permanent residents and businesses that were growing fast. Large numbers of boats unloaded goods at its Mississippi River docks and massive numbers of travelers came directly through that town. In contrast, Arkansas had no comparable large hub of commerce.

John Walker believed his land value would be considerably higher if it belonged to Missouri rather than Arkansas, so in a very bold move he talked to members of Congress before they voted on statehood for Missouri. He was able to convince them to redraw the lines so that his land was now indeed a part of Missouri.

The unusual bootheel is simply a line drawn around his property so that it would be included in Missouri when they joined the Union in 1821. Backroom deals happened in Congress even in the early 1800s.

The main part of the southern border of Missouri, set at the 36°30' line turned out to be a border that changed the country. The era was pre-Civil War, and as the nation expanded with the addition of the Louisiana Purchase, the tenuous balance of slave states to non-slave states threatened to tip. Until this point, under the Congressional order that no *new* states could practice slavery, the number of slave and non-slave states was about equal. When the massive new territory was added, the notion that it would be divided into individual states had slave proponents worried.

If all the new states were mandated as non-slave, pro-slavery voices would soon be greatly outnumbered in our representative government.

To settle the issue, legislators enacted an agreement known as the "Missouri Compromise" in 1820. The new state of Missouri could have slaves, but to counterbalance that, another new state that had just joined the Union (Maine) was *not* allowed to have slaves. From that point forward, Missouri's southern border of 36°30' was designated as the breakpoint for slavery. States to the north of that line could not have slaves; states to the south could. This would later have a tremendous effect on Texas (see chapter 28).

While the southern border of Missouri gets the nod as one of the most important historical borders because of slavery, the northern border managed to cause its own angst that lasted for decades. In 1816, surveyor John Sullivan didn't quite fulfill his charge to demarcate a straight-line border between Missouri and Iowa. Unintentionally, his surveyed line veered northward as it met its eastern point. Iowa raised an objection because the new border dug substantially into their territory, so they brought the issue to the federal government.

Eventually Congress hired a new surveyor who mapped a nice-and-straight new line. Unfortunately for Iowa, he drew it straight west from the point in the east that had already been too far north as a result of Mr. Sullivan's error. The completed line took over 2600 square miles of land from Iowa.

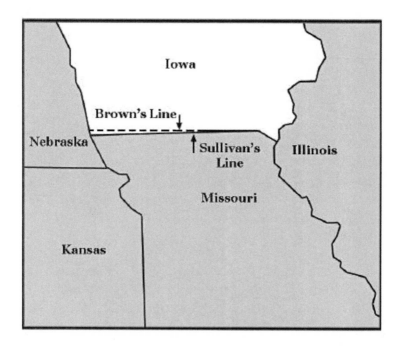

By the time an appeal made it to the Supreme Court in 1849, more than 30 years had passed and the judges ruled the line would stand simply because it had been that way for too long. Land had been deeded, people had settled, and the Court deemed it too disruptive to fix the mistake.

Chapter 25: Arkansas

Twenty-fifth state, admitted June 15, 1836

Arkansas began life as a basic rectangle that later had two notches knocked from it. When Congress started to divide the Louisiana Purchase, two historic decisions and a presidential mistake played a part in the shape of Arkansas we observe today.

The original design of the border between Arkansas and Missouri was the same straight line that created the northern border of Tennessee. But a funny thing happened on the way to statehood. A wealthy cattleman whose ranch covered a large portion of what is now the bootheel of Missouri watched with envy the thriving gateway city that St. Louis had become. He knew if his land were part of Missouri rather than Arkansas, that property would have a much higher value. He contacted members of Congress and made an interesting backroom deal.

In essence, he persuaded Congress to create the notch known as the Missouri bootheel so he and his land holdings could personally benefit from the commerce, and thus the wealth, which came through St. Louis. His gain was Arkansas' loss.

On the opposite corner of Arkansas, another chunk seems to be missing. The southern border of Arkansas was already set at the 33rd parallel when Louisiana entered the Union as a state in 1812, but no documents indicated the notch in Arkansas' southwest corner. What happened?

Arkansas was part of the Louisiana Purchase of 1803, but Spain owned most of the land to the west. Unfortunately the purchase agreement was unclear exactly where one tract of land ended and another began. When it came time to establish exactly who owned what and where the lines should be drawn, the Red River became an important bargaining point. In a decision called the Adams-Onis Treaty—the Adams in question was then Secretary of State John Quincy Adams—the Spanish territory would follow the Red River until it reached 88° longitude, at which point it would turn straight south. The result was a notch cut from the southwestern corner of Arkansas.

But a mistake by President Andrew Jackson finally solidified the notch. The line could have continued straight down instead of turning at the Red River. There was no real need for the notch to cut into southwestern Arkansas, but Jackson had made an error in some negotiations he conducted with the Choctaw Indians in 1820. With wording that was a bit vague, and included a misunderstanding of distances on the ground, the final borders of Arkansas were set in such a way that left the notch cut from its territory.

Knowing how Arkansas arrived at its shape is only one part of their story. Why don't they pronounce the name of their state Ar KANsas to keep in sync with their neighbor to the north? Both Kansas and Arkansas find the root of their names in a local Sioux tribe called The Kansa, but French explorers strongly influenced the settlement of Arkansas. Their influence extended to the pronunciation of the name they chose, and was notable in the silent "s" at the end. In an act of the state legislature in 1881 they declared the correct pronunciation was indeed "ArkanSAW."

Chapter 26:
Michigan

Twenty-sixth state, admitted January 26, 1837

B efore Michigan became a state in 1837, the territory was part of a larger landmass known as The Northwest Territory. From the map, you can see The Northwest Territory was made up of areas that later became Illinois, Indiana, Ohio, Michigan, Wisconsin and part of Minnesota. A Congressional document called the Northwest Ordinance* established the initial borders and mandated a territory had to achieve a population of more than 60,000 residents before it could apply for statehood.

Northwest Territory

The shape of Michigan is interesting because no other state is comprised of two peninsulas. The Lower Peninsula is nicknamed "The Mitten" because of the obvious shape, and the Upper Peninsula (UP) is not even connected to the rest of the state, but rather is attached to Wisconsin.

Why was the decision made to grab a piece of Wisconsin and then assign it to Michigan?

The Upper Peninsula story actually starts with Michigan's *southern* border. In the original wording of the Northwest Ordinance, the southern border of Michigan was a line drawn from the southernmost point of Lake Michigan straight east to Lake Erie. Settlers, seeking land for farming and new businesses, began to fill The Northwest Territory. Toledo quickly became established as a major port city on Lake Erie.

The borderline, as originally written, cut Toledo from the Ohio Territory. Port cities were crucial to a territory's trade and commercial development, and the pending enforcement of the borderlines made Ohio's territorial government nervous. They had invested heavily in Toledo, and depended on its port for much of their commerce.

Because the Ohio Territory had a greater population than Michigan, they were able to lobby Congress to redraw the line slightly to the north so Toledo would be included within Ohio's boundaries. Borders were a serious matter and the intervention of Congress prevented a conflict that was heating up. Referred to as the "Toledo War," Michigan and Ohio militias both organized to fight for the area known as the "Toledo Strip." However, because of Congress' intervention, no gunfire was exchanged and the borderline moved slightly to the north.

And then Indiana weighed in. Had the Northwest Ordinance's original line remained as described, Indiana's access to the Great Lakes would have been just a tiny point at the southernmost end of Lake Michigan. Again, because port access was so important, Indiana also lobbied Congress for relief.

As a result, the border between Michigan and Indiana was moved 10 miles farther north, resulting in another strip of land being taken from the territory of Michigan.

Since the modifications for Ohio and Indiana were done at two different times after two separate accommodations had been reached, the resulting line isn't completely straight, thus the little jog in Michigan's southern border.

In order to appease Michigan, unhappy its southern border had been nibbled away, Congress awarded the state the Upper Peninsula as compensation. Wisconsin objected, of course, because the UP was a part of *their* territory, but Wisconsin lacked the population threshold of 60,000 and were unable to mount a strong-enough campaign. The reality was: they had no vote in the matter.

Michigan's eastern border was critical because it didn't just define the boundary between two states; it also established a boundary between the United States and Canada. This border is actually extremely old, established in the 1783 Treaty of Paris that ended the Revolutionary War. As a result of the treaty, the eastern border of Michigan was defined as the edge of Lake Erie north to the southernmost point of Lake Huron, both lines bisecting the lakes. By using the Great Lakes as the state's boundary, both Canada and the United States would have access to the lakes for important transportation and commercial use. The boundary essentially splits both Lake Erie and Lake Huron between the United States and Canada.

> * The Northwest Ordinance was an act of Congress that established The Northwest Territory. This Congressional document was particularly important because it set the precedent that our westward expansion would establish new states rather than just expand the ones that already existed.

Chapter 27: Florida

Twenty-seventh state, admitted March 3, 1845

T he biggest segments of Florida's borders are quite obviously defined by the ocean, but even those coastal borders underwent drastic changes in Florida's history. About two-and-a-half million years ago, the landmass of the Florida peninsula was approximately twice the size we recognize today. When glaciers melted and the sea level rose, the exposed landmass shrunk to what 16th-century explorers found, and what we identify on modern maps. Ancient indigenous communities on the coast left behind artifacts and remains that are now under deep ocean waters.

Florida really has only one border that was established by decisions made by American statesmen, but even its northern border underwent several modifications before taking its current shape.

At the time the Spanish held ownership of Florida, their claim extended all the way west to the Mississippi River and much farther north into present-day Georgia and Alabama, making the panhandle a much longer and wider piece of real estate.

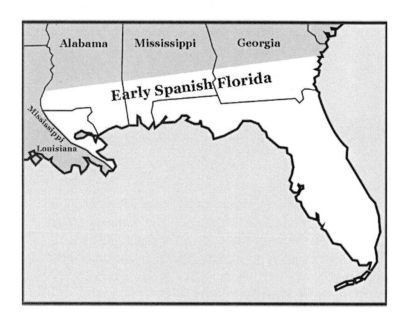

A series of bloody wars between the British and the Spanish picked at this extension of land, changing its overall size. A pact in 1739 legislated the St. Marys River would define the Florida/Georgia border.

But as most rivers do, the St. Marys River meandered independently through the inland region, and thus the border takes dramatic turns to the south and then back again to the north until it reaches the Okefenokee Swamp, creating the sizeable dip between territories. At that point all parties agreed drawing a straight line west until they met the Chattahoochee

River would make things easier. The swamp was not a place anyone wanted to settle, so the dispute was merely a technicality, not a fight over precious land.

At the Chattahoochee River, the border jumps north for 20 miles instead of continuing in a straight line. This line, at the 31st parallel, was left over from the treaty that established the Carolina Colony back in 1663. During the formation of Florida's borders, the territory changed hands twice: from Spanish to British and then back to Spanish again. Spain fought to extend their ownership north of the 31st parallel, but the new United States fought back and won the decision in the Treaty of San Lorenzo in 1795.

Later, the United States annexed other pieces of the long panhandle as a result of the famous Louisiana Purchase. In 1810, the United States claimed the chunk between the Mississippi River and the Pearl River. Three years later, the U.S. acquired the next piece from Spain, all the way to the current border formed by the Perdido River. In 1821, Spain left the territory altogether and released its holding of Florida to the United States with the perimeters as we know it today.

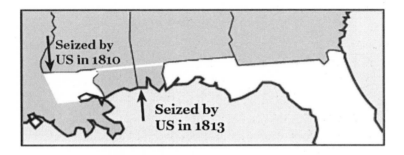

Chapter 28: Texas

Twenty-eighth state, admitted December 29, 1845

One of the most remarkable things about Texas is its size. But when Texas applied to be accepted into the Union as a state, the territory encompassed an even larger landmass than we know today. The issue of slavery was at the heart of why Texas (reluctantly) chose to reduce its footprint.

Much of Texas' east and northeast borders had been decided in a compact between Spain and the United States in 1819, when Texas was still a Spanish holding. The wording in the Louisiana Purchase was too vague to define clearly where United States ownership began and Spanish territory ended, so the Americans and the Spanish forged a treaty to settle the matter. The Adams-Onis Treaty (U.S. Secretary of State John Quincy Adams and Spanish envoy Lord Don Luis de Onis) established the Sabine River as the eastern border and the Red River in the northeast.

Things were just getting started with Texas. By that point, Texas looked like this:

Negotiations continued between Adams and Onis, but before much was settled, Mexico won its independence from Spain, making Onis' involvement obsolete. Now Adams shifted his arbitration skills to deal with his Mexican counterparts, who were wary of more and more American settlements in their treasured province of Texas.

In spite of Mexico's concern, over the years many more Americans settled in the region and eventually, in 1836, Sam Houston led Texas to independence from Mexico.

Being an independent republic, bordered on one side by Mexico and on all other sides by the union of states that comprised America, was a mixed bag of fun and folly. Big enough to be very powerful, Texas also found itself stretched to cover the costs of their war of independence and the governance of such a large territory. Texas' debts grew substantially, so ten years later they applied to be accepted into the Union.

Their timing was tough. Joining the United States meant they were going to have to comply with the Missouri Compromise, signed in 1820. The famous compromise established only states that fell below 36° 30' were allowed to be slave-owning states, but large chunks of Texas territory extended well above that latitude. They were faced with big decisions. Do they lop off massive pieces of valuable land in order to maintain their slave-owning culture, or do they disavow their past and go forward as a non-slave state?

The prospect of maintaining their economic future without benefit of slave labor was so frightening they chose to reduce the size of their territory. The result of that decision is the northernmost point of the Texas panhandle falls exactly at the 36° 30' mark. The territories of Kansas and Colorado absorbed most of the land they relinquished.

To the south, the Rio Grande River made a natural border between Mexico and Texas because it afforded both sides a degree of security, but that didn't become the official border until the Mexican War ended in 1848.

At this point, Texas still owned a large part of land farther west than today's map reflects. The same crushing debt that persuaded Texas to join the Union in the first place convinced them to sell more of their land to the west, an area that later became New Mexico. Congress also urged them to subdivide into five states but the suggestion wasn't received with much enthusiasm from Texas. Much like the situation with California (see the chapter on California), Congress felt they could push only so far. They believed it was better to have an oversized state as part of the Union than an independent republic within its own boundaries. Texas voluntarily reduced its own size in order to maintain slavery and reduce its debt, but would only go so far to satisfy the Jeffersonian ideal to have all states be roughly equal in size.

Congress also had an internal reason not to press Texas into dividing into smaller states. At that time, pre-Civil War, the number of slave-owning states was equal to the number of non-slave-owning states. Votes and voices were equal in Congress on the issue. Had they divided Texas into four states, slave-owning states would have attained a greater representation in Congress.

Chapter 29: Iowa

Twenty-ninth state, admitted December 28, 1846

I owa became part of the Union as a piece of the Louisiana Purchase in 1803. Congress quickly defined the eastern border as the Mississippi River, but the other borders were much harder to settle. In the early days of these "unorganized territories," the federal government removed Native tribes to make room for white settlers, and dickered with territorial governments over where to draw the lines for each new state. As had been the preferred practice, rivers were foremost in the bargaining chips, but the path to deciding on Iowa's borders was still a rough one.

Twenty-five years earlier, Missouri settled its own borders with the help of a Supreme Court ruling, but in doing so, created a serious issue with Iowa's southern border. Surveying errors, both in the original survey by John Sullivan and in a second one by Joseph Brown who was ordered to fix the errors in the first one, resulted in Iowa losing 2,600 square miles to Missouri. In spite of Iowa's complaints, the line stood that way for 25 years with Congress not willing to change it.

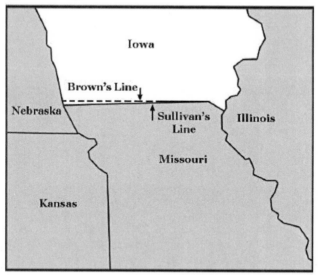

In Iowa's southeast corner, both sides *did* agree on one issue. Instead of following the line straight all the way to the eastern border, Iowa's boundary would dip to follow a small stretch of the Des Moines River, thus creating the little notch on the eastern edge of the southern border. Had they carried the line straight across, Missouri would have acquired an "island" on the other side of the Des Moines River that would have been awkward to govern.

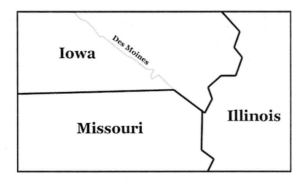

The western and northern borders of Iowa required extensive trade-offs between the territory and Congress. Congress preferred a narrower and taller state, mirroring the shapes they had sculpted into Illinois, Indiana, Michigan and Wisconsin. This taller configuration would have spread Iowa as far north as the Minnesota River, bringing it close to the location of modern-day Minneapolis. This version of the composition of Iowa also would have had a western border follow the straight-line portion of the lower part of Missouri's western border.

Iowa preferred a western border that stretched all the way to the line created by the *top* of Missouri's border, where the Missouri River angles westward. In the end they got their wish. To keep a more rectangular shape, the top portion of this western border switched to follow the Big Sioux River because the Missouri River took a sharp turn to the west. Iowa did *not* get their wish, however, when it came to their northern border. They liked the concept of claiming land all the way north to the Minnesota River, but Congress had a different notion. Now that Iowa's boundaries were stretched farther west than they originally envisioned, a lower northern border would keep the new state in line with the concept of creating states of nearly equal size. Setting the northern limit at 43° 30' established Iowa's height of exactly three degrees, a delineation that would see itself repeat when North Dakota, South Dakota, Nebraska and Kansas would be formed.

Chapter 30:
Wisconsin

Thirtieth state, admitted May 29, 1848

Wisconsin was the last of five states to be created from a large parcel of land called The Northwest Territory, won from France in the French and Indian War.

In an earlier decision, Congress had awarded a large portion of northern Wisconsin to Michigan—which came to be known as The Upper Peninsula—because of disputes Michigan had waged with its other border states. Despite objections from the Wisconsin territorial government, this section of land became part of Michigan to compensate them for ceding land to Ohio and Indiana on their southern border.

Wisconsin was still able to fight for a portion of that peninsula. Michigan claimed its jurisdiction should extend all the way to the western branch of the Montreal River, while Wisconsin argued Michigan had been given enough. From Wisconsin's point of view, Michigan's claim would stop at the eastern branch. Even after lengthy negotiations, no official proclamation was issued, but Wisconsin's version of events prevailed, and today that remains the division between the states.

Wisconsin had already lost another sizeable chunk of land to Illinois in the south. Congress ruled in favor of Illinois' appeal for over 60 miles of flat land in the southern part of the Wisconsin Territory so they could build a series of canals to improve their flow of commerce, specifically with the goal of reaching the Erie Canal. The discussion in Congress hinged on the arguments of North versus South and the anxiety of keeping northern business interests better able to move their products freely. Congress granted Illinois the land to build canals specifically for the purpose of easing business ventures in the North.

One last border had to be decided for Wisconsin: their western boundary. The original Northwest Territory drawings indicated this perimeter extending all the way along the Mississippi River, which would have cut deeply into the future state of Minnesota. Knowing that Minnesota would be applying for statehood shortly (ten years later), Congress had an interest in making sure the new neighboring state would have equal access to the Great Lakes. The original western border of Wisconsin would have prevented such access.

Congress adjusted the border to break north at the juncture of the Mississippi and St. Croix Rivers, thus creating a line drawn in a more northern path than the previous northwest slant. This new line ensured that both states would have ample access to Lake Superior.

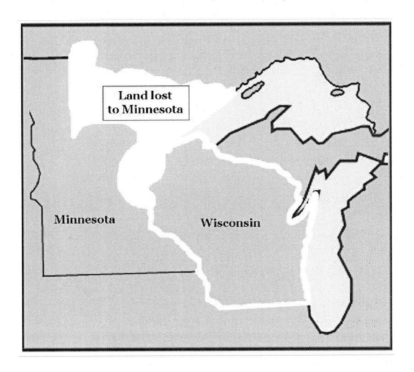

Chapter 31: California

Thirty-first state, admitted September 9, 1850

I f Thomas Jefferson made such a big stink about the importance of all the states being roughly the same size, how did California end up being so large? Congress could have easily created three or four states out of that landmass, so why didn't they?

When the Mexican War ended in 1848, California became an independent republic with a lot of overlapping ties to the U.S. territories. California had an obvious incentive to join the Union, because with such status came a larger structure for defense, the exchange of commerce, and camaraderie with other areas of settlement. The territory was certainly big enough to be a stand-alone nation, but it would be an isolated republic, butted on two sides by an increasingly powerful union of states with a military and an extremely large population.

The other entity, the United States Congress, also had
to digest these two scenarios. The Union was still in a
tentative status, and none of the legislators felt
absolute certainty the territories would maintain their
loyalty if times became tough. The potential existed
for one or more of the states to take a stand and break
from the Union at any time if they felt they would
have a better shot as a stand-alone country. Congress
had to consider the danger of having an independent
nation as large as California on our coast. Who could
predict years down the line whether that country
would be friend or foe?

Congress issued an invitation to California to join the
Union, although they issued it with a caveat. The
territory could join, but Congress would divide them
into at least three different states.

California understandably balked at this notion.
Strength comes in size and numbers, and left as is,
California had considerable strength. But was it
willing to press the point? Was Congress?

While both sides danced around the issue, something
happened that ended the debate. Less than a year
after the Mexican War ended, James Marshall
discovered gold near John Sutter's Mill and the Gold
Rush was on. With generous amounts of gold came
enormous power, and now California held all the
cards. Congress could keep pushing the issue of
dividing the territory, but they would do so at great
risk.

In the end, the U.S. Congress did not alter the size of
California. The eastern border was drawn along the

outline of the mountain range but, unlike mountain states to the east where borders were identified as the *ridge* along the highest points of the hills, California insisted their border be drawn on the *outside* edge of the mountains, ensuring that all the gold would be within the new, still very large state.

Congress originally pursued including the Baja strip to the south as part of this new state, but Mexico held firm. They would keep all lands south of the Gila River. The land itself held little value, but Mexico feared the U.S.'s ability to launch warfare attacks from their west. The final negotiation involved San Diego. The United States desired the important port city, and eventually both sides agreed to draw the line just south of San Diego, maintaining it as an American city.

Chapter 32: Minnesota

Thirty-second state, admitted May 11, 1858

Minnesota's unusual topknot resulted from an incorrect assumption made by the Congress when they finalized the U.S. border with Canada—then known as British North America. East of the Mississippi River, the border that divided our two nations was erratic and uneven, having been decided in bits and pieces as one by one, each territory became a state. As we sought to better define the demarcation between the United States and Canada farther west, both sides agreed just to establish a straight line, and use it all the way to the Pacific Ocean. The line they chose was the 49th parallel, a termination that ensured both the United States and British North America complete access to the valuable Great Lakes. Shipping and transportation on these waterways were crucial to the economic health of both nations.

After lengthy discussion and consultation, and with the treaty in hand, surveyors began marking the new line at a western notch of Lake Superior and then, by agreement, had to work their way up to the 49th parallel. They used a chain of lakes as their guide, which is how the northeast border of Minnesota achieved that series of zigzag lines. The problem was that each party to the treaty assumed The Lake of the Woods topped out at the 49th parallel. In the original wording of the treaty, a line through the chain of lakes was expected to end at the northeast corner of The Lake of the Woods, but it didn't.

When the surveying crew actually began the task of physically measuring the line, they discovered The Lake of the Woods extended much farther north than anyone thought. Given the choice of changing the charter and cutting the lake in half, or just redrawing the line, the decision makers opted to redraw the line to encompass the lake, thus creating the bumpy topknot on the head of Minnesota, which is officially called the Northwest Angle.

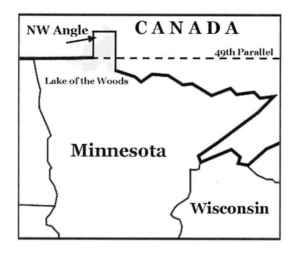

Minnesota was a state formed out of two separate territorial acquisitions. The eastern half was part of The Northwest Territory that became U.S. property after the French and Indian War in 1763. The western half was part of the Louisiana Purchase in 1803. When Wisconsin gained their statehood in 1848, Minnesota's interests had already been considered. The delineation between the two territories, under the boundaries formed from The Northwest Territory, was supposed to be the Mississippi River until it reached the northern portion of Minnesota Territory, but that separation would have left Minnesota without access to Lake Superior.

In accordance with Congress' insistence that all states have access to large waterways, they redesigned the border between the states. The new boundary began in the south, using the Mississippi River as previously described, but where the Mississippi and St. Croix rivers met, the line then followed the St. Croix River north to Lake Superior.

To create Minnesota's western border, Congress again chose rivers as the demarcation points. The Red River of the North accounts for about 200 miles of Minnesota's western border until the river takes a sharp turn east. At that point, the border picks up the Bois de Sioux River, and then later the Big Stone Lake creates the line. Only then did a conflict arise about the rest of the perimeter as it headed south to meet Iowa.

Minnesota argued the Big Sioux River should have defined the remainder of the border, and that the southwest corner should have met Iowa's northwest corner. The problem turned out to be budget. Congress hadn't yet surveyed this part of the country, and didn't have the budget or manpower to do so at that time. The simpler solution meant drawing a straight line from the last point with which they were familiar, and this remains the southern stretch of the western border to this day.

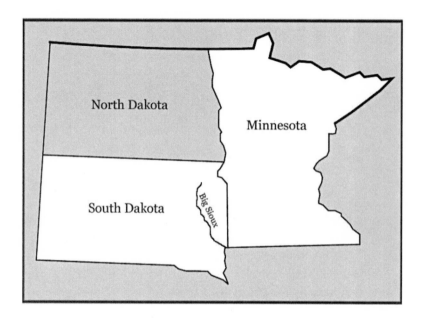

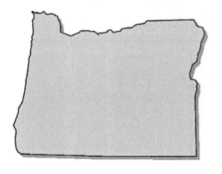

Chapter 33: Oregon

Thirty-third state, admitted February 14, 1859

B y the time Americans decided to settle the territory that would become Oregon State, its own indigenous people had already thoroughly explored the region, as had explorers from Spain, Russia and England. After the Louisiana Purchase in 1803, traders and settlers quickly moved into the territory and expanded their settlements all the way to the northern Pacific Coast.

The southern border was the first to be settled because negotiations with Spain had already defined it. In 1790, England and Spain divided the region with a treaty promising England could have the territory north of the 42nd parallel, with Spain in control of lands south of the area. This demarcation was a good, natural dividing line because of what it had to do with the waterways, the all-important resource settlers used to move people, goods and animals. Rivers north of 42° emptied into the Columbia River and its water, in turn, flowed into the Pacific Ocean at modern-day

Portland. Waterways to the south of 42° flowed either directly into the Pacific or meandered all the way to San Francisco Bay. In this way, both nations could enjoy access to means of moving commerce without having to risk the potential hurdles of traveling through another country's territory.

The northern and eastern boundaries were not quite as simple. As was the case in so many other negotiations to define a state's shape, Oregon began its life as part of a much larger territory. The region, called "Oregon Country" was later divided into Oregon, Washington, Idaho, British Columbia and parts of Montana, Wyoming and Alberta.

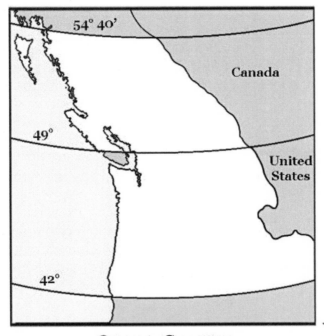

Oregon Country

When lawmakers made the decision to divide the portion south of the 49th parallel—the recently established border with Canada— Congress created a new Washington Territory. In order to define Oregon, they identified the Columbia River as the divider, beginning on the Pacific Coast all the way to the point where the river turns north and crosses the 46th parallel. At that point, the line headed straight east, all the way to the Continental Divide. Much later we would see a scaling back of the eastern expanse of Oregon.

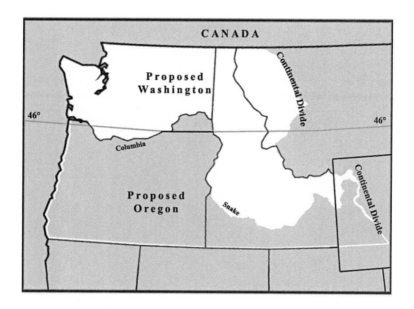

Using the Columbia River as the starting point was not a random decision. The important port where the river converged with the ocean was crucial to Oregon's economic development, and they knew the future Washington State would have its own substantial port in a city later called Seattle.

At the time Oregon achieved statehood, the new state government voluntarily released some of its eastern territory. Previously extending all the way to the Continental Divide, they now agreed to use the Snake River as its line of demarcation. The new border ran south along the Snake until it joined the Owyhee River about halfway, and at that point they drew a straight line south to meet the southern border. By cutting the size of Oregon back as they did, they met the ideal of creating a state with exactly seven degrees of width. Using that same eastern perimeter, Washington would be granted the same span. In the end, these two states joined Wyoming, Colorado, North Dakota and South Dakota with widths of seven degrees. Once again the planners kept Jefferson's ideal of states of the same size, or as close as they could get, in the forefront while they sliced up the landmass.

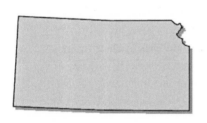

Chapter 34: Kansas

Thirty-fourth state, admitted January 29, 1861

Most of what is now Kansas emerged from the larger Louisiana Purchase of 1803. Kansas' eastern border had been determined when Missouri achieved statehood in 1821, but as Congress began the process of slicing the territory we acquired in the Louisiana Purchase, Kansas was a much longer and wider entity. The decisions made about Kansas' three other borders hinged on the outcome of the fight over where American landowners could continue to own slaves.

One of the urgencies in creating a governed state of Kansas were the failed attempts to run a railroad west from Missouri to the Rocky Mountains of Colorado. Without a legitimate state with a governing body, the vast areas of land in the middle of the country were wild and dangerous. Stephen Douglas started the Illinois Central Railroad and he envisioned Chicago as a hub of railroad commerce. He recognized that running tracks through the new Louisiana Purchase territory was an important part of that railroad's development.

Spurred by railroad interests and the national government's own interest in turning the vast tract of land into governable communities, Congress made decisions about Kansas' north and south borders.

163

Many of those decisions were met with angry confrontations. People argued loudly about the line set by the Missouri Compromise, which declared states north of the 36° 30' latitude would not be slave-owning, while states to the south were allowed slaves. When Congress set Kansas' southern border at 37°, the decision affirmed the non-slave status of the new state, and also established a line that would later create Oklahoma's panhandle (see the chapter on Oklahoma). Planning ahead, the secondary focus for setting the boundary at 37° meant Congress would later be able to create four states of exactly three degrees of height between 37° and 49°, the United States' border with Canada.

Creating a state with three degrees of height meant the northern border of Kansas was firmly set at 40°. The only thing Congress had left to decide was how far west the new state would extend.

Three years earlier, in 1858, gold had been discovered in the far western portions of the vaguely described Kansas Territory. With gold came an invasion of 50,000 outsiders of various and not-always-honorable intentions, and the small territory suddenly found itself overwhelmed by crime and chaos. Concerned for the stability of their new state, Kansas divested itself of the gold mines to the west and set their own western border at 102° longitude.

Once again Jefferson's ideal of creating states of equal size played a large part in decisions made about the shape of one of our states. Wishing to create four states with exactly three degrees of height was a part of that design. By setting Kansas' western border at the 102nd longitude, they also created a state that was seven degrees of width, a dimension that repeated itself in many other states in the west.

Chapter 35: West Virginia

Thirty-fifth state, admitted June 20, 1863

How did West Virginia get that goofy shape? Is it possible someone intentionally created a state to look like this?

Not exactly. West Virginia was "spun off" from Virginia when residents on both sides of the Appalachian Mountains acknowledged they had little in common with each other and also, transportation between the two regions was extremely difficult. Land transportation in the 1800s was primarily horse and buggy, and a mountain range was a huge impediment to travel.

Virginia Before the Split

A disparity also existed in the availability of standard government protection and services. The governor never physically visited the western part of Virginia located over the mountains, and law enforcement agencies had difficulty carrying out their services there.

Additionally, the two populations on either side of the Appalachian mountain range were vastly different culturally. Coal mining heavily impacted west Virginia's commerce, and they suffered from very little arable land for agriculture. The eastern portion of the state had an economy spread between generous amounts of agriculture and a prosperous shipping industry. In addition, the poorer residents to the west did not have the economic capacity to own slaves as they did in the larger, eastern portion. This then became an issue because Virginia's legislative representation used census numbers in which slaves were counted—but not, by extension, given the right to vote. Poorer, less populous western Virginia had very little voice in policies debated in the state legislature.

With those factors in mind, West Virginia voted to secede from Virginia at the same time Virginia seceded from the Union at the onset of the Civil War. West Virginia fought for the Union, while Virginia sided with the Confederacy.

The meandering lines of the Appalachians essentially defined the border between Virginia and West Virginia, but the border was also influenced by Congress' interest in making West Virginia a viable state. Because West Virginia had poor growing land and fewer natural resources, Congress seized a piece of rich property from the Virginia side and added it to West Virginia's domain. The odd drop-down notch at the northeast corner of the border between Virginia and West Virginia (circled on the map below) was an addition Congress made to West Virginia's territory because the fertile land would provide a means for West Virginia to become self-sustaining. In addition, this land contained an important railroad line the Union naturally wanted to keep out of Confederate hands. Virginia argued vehemently to keep that land, but having recently seceded from the Union, Virginia had virtually no bargaining power with the U.S. Congress.

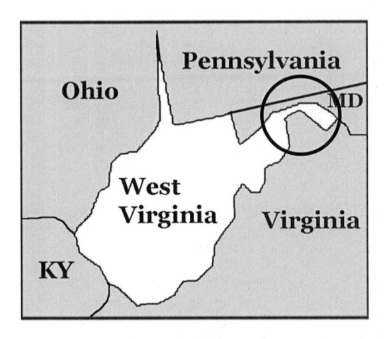

The little finger of West Virginia that snakes up the side of the Pennsylvania border was left over from an earlier dispute Virginia and Pennsylvania had over another piece of territory, the place where the Monongahela and Allegheny Rivers met. Both territories laid claim to this land, and when Congress awarded it to Pennsylvania (who called the new city there "Pittsburgh"), Virginia received compensation with a swatch of land to the west. When West Virginia was created, that finger of land along the Ohio River became part of the new state.

Chapter 36: Nevada

Thirty-sixth state, admitted October 31, 1864

Why does Nevada come to a point at the bottom? The state didn't always look that way. The earliest version of Nevada showed boundary lines in a more rectangular shape, with a sideways slant to accommodate the angled border of California.

Who looked at the map and said, "What this state needs is a point"?

Nevada emerged from the much larger area called the Utah Territory at about the time prospectors discovered silver and gold in the mountains. Precious metal discoveries elsewhere in the region had led to a massive influx of people, and the new Nevada territory was no exception. When the Civil War began in 1861, the North needed the metal resources in even

171

greater quantities to pay war expenses and for use in making weapons.

During the Civil War years, the boundaries that defined the fledgling territory kept getting wider and wider as gold and silver strikes kept spreading. Congress had an additional interest in expanding Nevada because the more they gave to Nevada, the less land would be assigned to Utah, a rebellious territory led by Mormons who insisted on governing their land as a religious body, rejecting interference by Congress. Congress answered their demands by dividing the region and giving land to Nevada.

Eventually Congress also wanted the desert territory to have an access to a substantial waterway. Land commissioners widened the boundaries a third time to include a piece of the Colorado River. But the Nevada Territory still didn't have a point at the bottom. What happened?

When Congress granted Nevada the area that is now the point, the plan was to give the new state even better access to rivers that would allow transportation all the way to the Gulf. Steamboat transportation on the Colorado River would become a very valuable asset to Nevada, providing means to transport goods to centers of commerce. This time, the land they claimed came from the Arizona Territory whose legislative body fired back an angry response. In spite of their organized and vehement answer, Arizona did not win that dispute. During the Civil War they aligned themselves with the Confederacy, and were not likely to gain any favors from the prevailing Union.

Nevada eventually stopped growing and kept their width and their point.

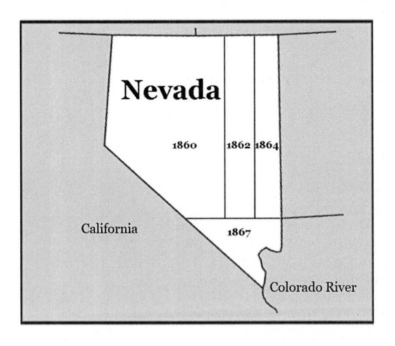

Chapter 37:
Nebraska

Thirty-seventh state, admitted March 1, 1867

T he state of Nebraska emerged from a much larger Nebraska Territory that in turn, began life as part of the gigantic Louisiana Purchase. Nebraskans never expected the final state borders would be as large as the territory started out, and much of what eventually defined the proposed state had already been decided by other states.

When the neighboring states of Iowa and Missouri attained statehood, Nebraska's eastern border was established at the Missouri River. Similarly, Nebraska's southern and northern borders were predetermined when Kansas' borders were finalized. Congress long promoted the ideal that all states should be as close to the same size as possible and, although many factors prevented any kind of grid-like division of territory across the country, when they had the opportunity to divide the Louisiana Purchase, that idealistic division of land actually had a wisp of possibility.

When Congress set Kansas' southern border at 37° through the Kansas-Nebraska Act of 1854, they set in motion a plan to create four states stacked on top of each other with exactly three degrees of height. The distance between the 37th parallel and the 49th, the border with Canada, allowed for exactly four states with the same height to follow. Those states would become Kansas, Nebraska, South Dakota and North Dakota. With that in mind, Nebraska's southern border was the already-existing northern border of Kansas (the 40th parallel), which in turn automatically defined Nebraska's northern border as the 43rd parallel. Nice straight lines, drawn with an allowance for the zig and zag of the Missouri River on Nebraska's northeastern corner, completed the mapping of those two borders.

But what about the large notch in the southwest corner of Nebraska? Nebraska earned ratification as a state before Colorado, so it seems Nebraska should have had first dibs on that chunk of territory, and, in fact, they did. They willingly gave it to Colorado. Why would they do that?

Nebraska's western border had been set at the 104th meridian because that made the state seven degrees in width, one of the uniform ideals that Congress emphasized. Eventually the states of North Dakota, South Dakota, Wyoming, Colorado, Nebraska, Washington and Oregon would *all* span seven degrees in width. The 104th meridian line was originally extended all the way south to Nebraska's southern border at the 40th parallel, but relatively quickly Nebraska carved out a notch and gave it to Colorado. Two important factors impacted this decision.

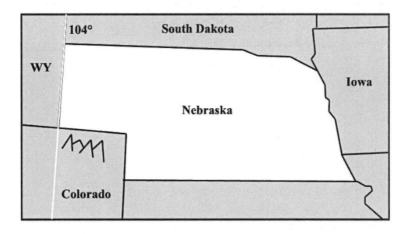

The low-lying Platte Valley defined large parts of Nebraska, and Territorial Governor Alvin Saunders was focused on attracting a great central railroad, an asset Kansas also coveted. This extensive transportation plan would serve to connect the Atlantic coastal states to the new settlements developing all the way west to the Pacific Coast. A railroad was the perfect industry for this flat, newbie state, and with a railroad running through, more commerce would surely follow. The southwest corner created two headaches for that mission.

First, mountains covered the area, which meant the land was not conducive to the planned cross-country railroad.

Second, again, mountains covered the area, and the recent discovery of gold meant miners would follow. Nebraskans had witnessed the chaos, crime and general disorder that ravaged other territories when tens of thousands of miners brought their culture and disorganized communities into a region. Not eager to

177

suffer the fallout from that type of upheaval, Nebraska happily turned that part of their land over to Colorado, thus creating the notch.

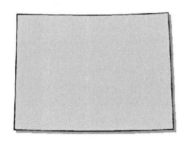

Chapter 38: Colorado

Thirty-eighth state, admitted August 1, 1876

A state whose shape is a neat rectangle may not seem to have any interesting historical explanations, but Colorado has a back story to explain why it has that shape, and also why it ended up in its modern-day location.

Colorado wasn't even imagined in 1850 when the huge Utah and New Mexico Territories originated. Nor was Colorado on the radar in 1854 when the Kansas-Nebraska Act set in motion the slicing up of part of the Louisiana Purchase into Kansas and Nebraska Territories. The state we recognize now as Colorado formed right in the middle of where all of those territories met, but its original name was "Jefferson."

When miners discovered gold in the Kansas Territory in 1858, that territorial government recognized it wouldn't be able to handle the influx of 50,000 miners and other gold seekers. Legislators worried about how the over-worked law enforcement officials would be able to manage the chaos.

In response, local residents held a special meeting a year later, and founded a new territory called *Jefferson* to encompass all the areas in which gold mining was creating havoc in the region. Initially, this proposed area was considerably larger than modern-day Colorado, and predictably the new borders immediately came under dispute.

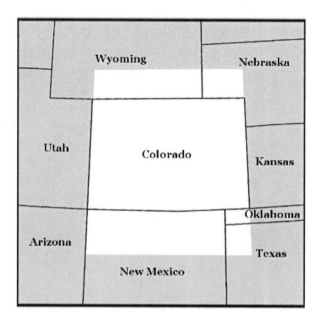

The east and west borders proposed by the residents of the newborn Jefferson met with no argument from Congress because it fell neatly into their ideal of having states closely match each other in size. By the time they were done cutting up the western half of the nation, North Dakota, South Dakota, Nebraska, Colorado, Wyoming, Washington and Oregon would all be seven degrees in width.

The greater height of Colorado, as it was renamed in 1861, did *not* sit well with Congress, especially when those boundaries encroached on two states that already existed: Nebraska and Texas. Locating the southern border at 35° also encroached on the recently acquired Santa Fe area for which Congress was already formulating other plans—a territory called New Mexico. Accordingly they raised the southern border to the 37th parallel and dropped the northern border from 42° to 41°. The result was a state exactly four degrees of height that satisfied another of the designer's ideal. Colorado was now a state whose height would later exactly match the heights of Wyoming and Montana, which were founded a little more than a decade later.

A final Enabling Act, passed on March 3, 1875, laid out all the specifications of the new rectangular state.

When women won the right to vote in Colorado in 1893, it became the first state in the Union to grant this right through a popular election.

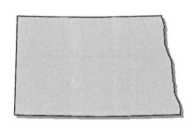

Chapter 39:
North Dakota

Thirty-ninth state, admitted November 2, 1889

The northern limit of North Dakota appeared in the 1850s when The United States and Canada agreed on the 49th parallel as the border between the two countries. But even that was a little muddled because we acquired only *part* of what became North Dakota in the Louisiana Purchase. The northeast corner of the state still belonged to France.

As it happened, the wording of the Louisiana Purchase left a few things open to interpretation and confusion. By the time Congress set out to design North Dakota, Minnesota had already been ratified as a state. So, if France hung onto that little chunk of property, a French island would have been created in the midst of American states, bordered by Canada to the north. Very little debate followed. France did not fight for the land. North Dakota absorbed the region in its entirety, and very quickly both the northern and eastern borders were accepted.

Congress had also already set in motion an ideology that the states they created out of the giant Louisiana Purchase should be seven degrees in width. For practical purposes, this didn't work for all the states, but because Nebraska had already been formed to the south, the western border for North Dakota was all but a done deal. For the first time, Congress got a two-for-one special. By continuing the western border of Nebraska all the way north to the border with Canada, and then cutting the newly created territory in half lengthwise, the government created both North and South Dakota at the same time. Both were three degrees in height, like their neighbors Nebraska and Kansas to the south, and seven degrees of width. The uniformity reflected Thomas Jefferson's great wish to create states close to the same size so they would enjoy equal representation and a hypothetical equal access to natural resources. See "Jefferson's Ideal of Equal States" on page 31 for a further explanation of his plan.

Late in the 20th century, a popular discussion pondered the question of why we didn't just have one large state named "Dakota." The answer lies in that ideal of creating states of equal size. The heights and widths of both Dakota states were set in stone before either state existed.

To the really picky observer, it is true the southern border of North Dakota is off by one-twelfth of a degree. Instead of the idealized 46° line, Congress agreed to shift it slightly south for the convenience of placing the borderline on a plateau. Sometimes practicality took precedence over precision.

Chapter 40:
South Dakota

Fortieth state, admitted November 2, 1889

As noted in the chapter on North Dakota, Congress created the two Dakota states at the same time and perhaps something as simple as a coin flip decided which would be designated the 39th and which the 40th state.

Minnesota's ratification in 1858 substantially defined South Dakota's eastern border. And, as it turned out, the state's northern, southern and western borders did not require much debate either. When the U.S. bought the Louisiana Purchase in 1803, Congress already planned to divide the immense territory into states that were as close in size to each other as possible. The ratification of Kansas in 1861 pretty much locked up the borders of what was to become South Dakota 28 years later.

Kansas' southern border at the 37th parallel had already been chosen specifically because that location meant the expanse of land between there and the 49th parallel, The United States' border with Canada, could be divided equally into four states with exactly three degrees in height. As Kansas, Nebraska, South Dakota and North Dakota were formed, the mapmakers had an easy job. With small exceptions, those boundaries became straight lines drawn every three degrees.

Uniformity also played a big part in determining the width of both the Dakota states. Desiring each Dakota state form a territory with seven degrees of width, Congress simply extended the line they already created for Nebraska's western border all the way to our border with Canada, thus locking down South Dakota's straight western border.

As simple as South Dakota's border formation seems, they did have a glitch in the southeast corner that didn't arise until much later.

A 20th-century disagreement between South Dakota and Nebraska was typical of the kinds of border problems Mother Nature creates. The winding route of the Missouri River defines a portion of the border that separates Nebraska and South Dakota.

For many years, the river boundary gave the two states no problem, and citizens on both sides co-existed without issue. But rivers don't always follow the same course year after year, and when floods eroded the banks, the force of moving water caused the tumultuous river to meander off course. In the 1950s, newly constructed dams settled things down temporarily.

But the gradual changes caused anxiety on both sides, and everything came to a head in 1977 when Congress granted the Missouri River National Recreational River status. Suddenly the spotlight was on fishing and waterfowl hunting laws that hadn't mattered before. South Dakota has a law called the "Duck Bill" that bans waterfowl hunting by non-residents. The problem was that over the course of decades, some of the land that Nebraskans had always used for hunting was now on the South Dakota side due to flooding and erosion.

The urgency to ratify a treaty and settle the hunting issue once and for all escalated. Eventually, a deal allowing a limited number of permits for Nebraskans to hunt on their old favorite spots settled that dispute.

Other erosion problems resulted in homeowners losing land and created legal issues that led to the higher courts. St. Helen Island used to be on the South Dakota side of the river, but flooding shifted it to the Nebraska side. At the same time, a parcel of land called Elk Island was no longer an island after a shift in the river left it attached to the South Dakota bank. Landowners still paying taxes to Nebraska found themselves *living in another state*. Do people go to bed as a Nebraskan and wake up as a South Dakotan?

The dispute continues and likely will not be satisfied soon. A Supreme Court ruling may be required. Part of the confusion is that one set of laws kicks in if a river changes course over a long time, because of erosion for example, and a different set of laws is activated when a change happens overnight, perhaps due to heavy storms or a spring melt that causes extensive flooding. Residents in this area await a decision from the court.

Chapter 41:
Montana

Forty-first state, admitted November 8, 1889

Montana was one of several states created from a larger landmass called the Idaho Territory.

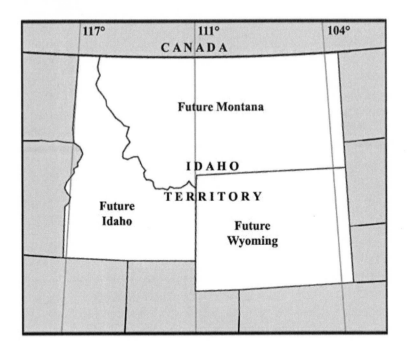

Because North and South Dakota were ratified ahead of Montana, the future Montana's eastern border was set at the 104th meridian. Her northern border was the boundary set by a treaty defining the border between the United States and Canada. Additionally, Montana's southern border, at least the straight-line portion of it, had been determined when Colorado was created 13 years earlier.

When Congress designed Colorado to be precisely four degrees in height, the location of those borders provided them another opportunity to map an area with states of approximately the same size. The distance between Colorado's northern border at the 41st parallel and the international border with Canada at the 49th parallel meant Congress could create two more states with the same four degrees of height that described Colorado. Wyoming and Montana were the two states that filled the eight degrees of separation.

At a glance, Montana's western border appears to betray all the planning that had gone into creating states of equal size. How did they decide upon a shape that extended so deeply into the future state of Idaho?

Much of this portion of the larger Idaho Territory was comprised of the middle section of the imposing Rocky Mountain Range. As discussed in earlier chapters about states that formed around The Appalachians, legislators knew governing a state with land on both sides of a mountain range was extremely difficult. Congress agreed that sculpting a state of Idaho, which was already in the planning stages, with a towering travel impediment due to the mountains, would saddle them with a burden they didn't need.

They chose, instead, to extend Montana's boundaries west to encompass most of that mountainous region.

However, they didn't cut Idaho off from *all* of the mountains, so how did they determine just where the line would fall? Political bickering played a large part in drawing this part of the map.

Idaho had argued for a bigger swath, much wider than the narrow neck we see on maps today, but unfortunately the territorial governor made the mistake of insulting the person who ended up having great influence over that boundary placement. Former Congressman, now Judge, Sidney Edgerton felt snubbed over an assignment by Idaho's governor that placed Edgerton's jurisdiction in a remote part of the territory. Edgerton believed the isolated outpost was an affront to his dignity and standing in the community. Unfortunately for Idaho, the former Congressman was a man with friends in high places and, in fact, he counted Abraham Lincoln among his friends. With influence from Washington, D.C., Edgerton was able to get a boundary line drawn much more in Montana's favor than Idaho's. Congress did pull back a little from Edgerton's initial recommendations because they argued his restrictive line would cut Idaho from owning some very fertile valley areas. They were intent on ensuring that Idaho would have some rich agricultural land. Montana already had plenty.

The bulge at the southwest corner of Montana's perimeter was a left-over from an earlier debate about the split between Montana and Idaho. One of the original suggestions for the border between the two states had been to follow the continental divide for that entire length, but Edgerton's interference ambushed that idea. A straight line, continuing Montana's border with Wyoming on the 45th parallel would have certainly been easier for mapmakers, but the dip to the south reflected a border that *did* follow the continental divide, at least for a small segment.

Chapter 42: Washington

Forty-second state, admitted November 11, 1889

Washington was one of three states that voluntarily turned down a fortune in gold. Previously Kansas and Nebraska each proactively surrendered chunks of their territory in order to protect themselves from the crime, chaos and influx of tens of thousands of enthusiastic gold seekers who strained a state's ability to enforce the peace.

In the mid-1800s the territory of Washington was considerably larger than the current state observed today. Oregon had just achieved statehood and established north and east borders that helped define the Washington Territory into something that looked like the white area below.

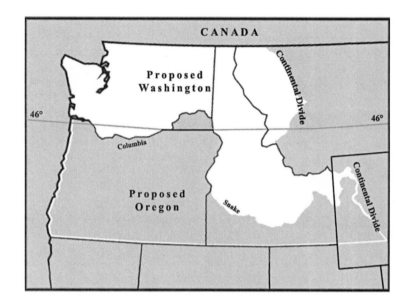

In 1860, the discovery of gold in the mountains in the eastern part of Washington Territory changed the plans for the future *state* of Washington. Instead of seeing gold as a fortuitous windfall that would enrich the entire region, the territorial government feared the worst. Just two years before, when miners discovered gold in Colorado, a tide of almost 50,000 people descended upon that area. The influx brought a level of lawlessness the territorial government of Colorado found extremely difficult to control.

Worrying about the potential unruly behavior of all these many miners was only part of the problem for the territory of Washington. Governing a wide area separated by mountains was a task in itself, but the cultural makeup of the type of people flooding the mountain areas was quite different from the settlers who were already in the region, based primarily near

Puget Sound. Fearing the large numbers of newcomers would soon have a political voice that could change the cultural landscape of the soon-to-be-state, lawmakers made the decision to divide the territory and cut off those "problem" mountains.

As it turned out, Congress was a step ahead. They had just completed the work on Montana's western border and had an eye on how to divide the land between Montana and the Pacific Coast. Oregon's eastern border had been set in 1859, and that perimeter gave a clear path to use the same line for Washington.

Starting at Oregon's northeast corner, Congress continued Oregon's border north of the 46th parallel, using that portion of the Snake River to begin Washington's eastern enclosure. Where the Snake met the Clearwater River, they switched to a straight line north all the way to the U.S./Canadian border. Once again Congress created a state that had seven degrees of width, along with Oregon, North Dakota, South Dakota, Colorado and, eventually, Wyoming. Relative equality of size continued to be of paramount importance as states took shape.

Washington's northern border required some diplomacy. In 1818, the 49th parallel became the demarcation between U.S. and Canada, but on the west coast, many people, including presidential candidate James K. Polk, insisted U.S. interest extended all the way north to the line at 54° 40'. Promoters of that ideal even coined a phrase "Fifty-Four Forty or Fight" to represent the intensity of the concept. But in the end, then-President Polk signed a treaty with Canada, accepting the 49th parallel as the border. The treaty also allowed British Canada to keep all of Vancouver Island, even though it extended south of 49°.

Chapter 43: Idaho

Forty-third state, admitted July 3, 1890

A ll this talk about creating states of essentially the same size, and then Congress dreams up Idaho. How did that shape that looked like a long neck happen?

The Idaho we know today was created out of the pieces left over from whittling a larger territory. Montana, Washington and Oregon already achieved ratification, so in 1890, the government took the leftover land and sculpted Idaho and Wyoming.

Idaho's southern border had been predestined since 1790. In that year, England and Spain ceased their bickering over access to waterways and agreed on the 42nd parallel as the dividing line between the lands each country held in that region. Later, in 1818, when British Canada and the U.S. identified the 49th parallel as the boundary between nations, Idaho's northern border existed even though Idaho itself didn't.

The east and west borders, which create the interesting long-neck shape that makes Idaho so distinctive, appeared much later.

At one point part of the future Idaho was the eastern chunk of what was called the Oregon Territory:

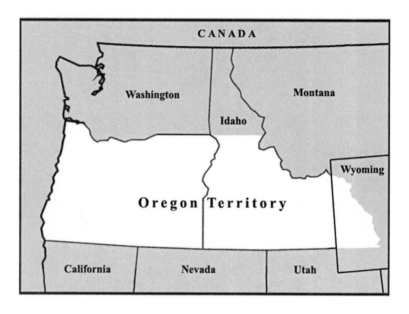

... but when Oregon state was created, another entity, Washington Territory, absorbed that chunk of land and more:

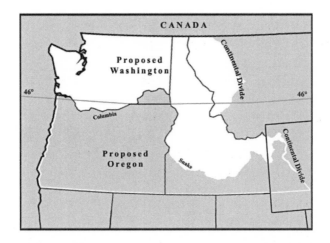

Washington Territory underwent its own change when Congress defined the final borders of both Montana and Washington State two days apart in 1889. This left a long neck and large hunk of a body for Idaho, along with a sizeable mass that would soon become Wyoming. Washington Territory chose to divest itself of the eastern portion of its expanse in large part because they feared what gold would do. They understood the difficulty in governing a region where the discovery of gold left them vulnerable to a large-scale infusion of new and often lawless adventurers. Mountains also separated their eastern from their western province, and this added to the difficulty of keeping the people in the eastern portion reasonably protected by law enforcement agencies that had trouble reaching the region. By dividing the territory where they did, Washington State wiped its hands clean of the resulting turbulence and law enforcement issues.

Congressional decisions made in 1889 meant Montana's western border bulged deeply into Idaho's

199

space. To better serve the largely mountainous region, Congress decided it should be under the governance of one state, rather than two. Idaho's territorial legislature argued the Continental divide made the most sense as a perimeter and would afford Idaho a larger, wider landmass. However, Idaho ran up against a former congressman with friends in high places, who had a bone to pick with the territorial governor of Idaho.

When Sidney Edgerton received a judicial appointment in Idaho, he accepted the honor as a former congressman, embracing the new title and the influence it would provide. When he discovered his appointment was in a remote district east of the Rockies, he elected to retaliate for what he believed was a humiliating professional snub. He used his clout with Congress, which was in the process of establishing Montana's western border and, relying on his friendship with Abraham Lincoln, pressured the government to reject Idaho's argument. He convinced them to push the dividing line between Idaho and Montana west to the crest of the Bitterroot Mountains, creating the much narrower neck that today is Idaho's northern half.

Congress did make a favorable concession when they created the straight line for the northernmost part of Idaho's eastern border. Had they followed the Bitterroot Mountain crest all the way to the 49th parallel, their decision would have deprived Idaho of the precious Kootenai River watershed agricultural land. One of many considerations Congress made when dividing the states was to assure each state access to a variety of natural resources, and Idaho needed arable land.

Chapter 44: Wyoming

Forty-fourth state, admitted July 10, 1890

B y the time westward expansion brought settlers across the Mississippi River, states started to take on relatively uninteresting and unimaginative shapes, like rectangular. As explained in the chapter called "Squiggly States in the East, Boxy States in the West" on page 21, the advent of railroads removed the need to use rivers to define the borders of states. Congress found it much easier to draw boxy shapes, and attempt to make states roughly the same size.

But Utah and Wyoming were both rectangular shapes that overlapped each other. One was going to have to give up some land. How did they decide to leave Wyoming as a complete rectangle and cut a notch in Utah?

Part of the reason was political. By this time Congress had encountered the Mormon influence that flooded the region and were not receptive to granting that fledgling territory any more land than necessary (see Chapter 45 on Utah for more information). The notched area contained many rich natural resources, and Congress worried about giving too many assets to a religious group that had been overly aggressive in demanding a political role in the administration of their state.

But in the end, more practical grounds guided the decision to grant that land to Wyoming. Just to the south and west of the notch, Utah is home to the rather sizeable Uinta Mountains. The ability to govern a parcel of land separated from everything else in the state by a wall of mountains was a serious difficulty in those times. From the Wyoming side, no such impediment existed. Granting that notch of land to Wyoming made pragmatic sense based on their geographic ease of maintaining law enforcement and state governance in that area.

So how did Congress decide the rest of Wyoming's borders? Mostly by default. Before Wyoming existed, North and South Dakota to the east already had a place on our maps, and Montana and Colorado's borders were set. This meant Wyoming's eastern, northern, and southern borders were agreed on, and the only one left to consider was their border with Idaho. Idaho and Wyoming would be ratified just seven days apart, and uniformity and simplicity were in the forefront of Congress' mind when deciding how to split them. By placing Wyoming's western border where they did, Congress created another state with exactly seven degrees in width, joining Washington, Oregon, North Dakota, South Dakota and Colorado. Rectangular Wyoming also shares uniformity with Montana and Colorado as three states with exactly four degrees in height.

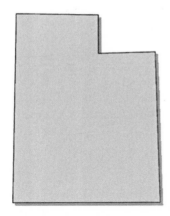

Chapter 45: Utah

Forty-fifth state, admitted January 4, 1896

U tah's history included a skirmish with the federal government years earlier over whether a religious organization could run a state government. This heated exchange came to a head when Mormons pushed for a sanctuary state called "Deseret."

When in 1847 Brigham Young led his Mormon followers westward from their settlement in Illinois, he envisioned a vast territory that would be a haven for the beleaguered Mormon followers who had been shunned in Illinois for their polygamous practices. At that time not much of the west had been officially divided, so Young and his followers envisioned their new territory to look like this:

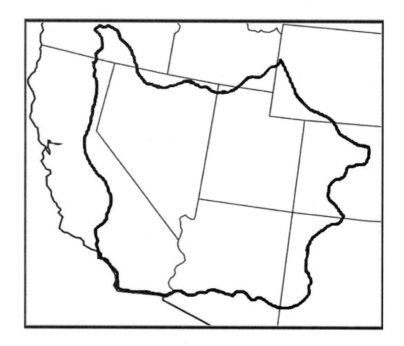

Congress had other ideas and eventually divided the area quite differently. For more on the conception and plans of Deseret, see "States that Didn't Make the Final Cut: Franklin and Deseret" starting on page 23.

By the time the United States prepared to accept Utah into its Union, many of the neighboring states had firm borders. Utah's northern limit had been set in stone as far back as 1819 when a treaty between the U.S. and Spain established the 42nd parallel as the dividing line between Spanish and American territories. At that time, Spain controlled the land south of the 42nd parallel, but in 1848, the U.S. seized that land at the end of the Mexican War.

When Utah gained admittance as the 45th state, neighbors Nevada, Idaho, Wyoming and Colorado all existed and had firm borders. Also, six years earlier, Congress decided in Wyoming's favor in their dispute with Utah over which jurisdiction would take a notch out of the other. Congress settled the matter when they concluded the mountainous area to the south and west of that notch created a difficult hurdle for a legislature in Utah to govern. Mountains were still a tough barrier when travel was limited to trains and horse-drawn wagons.

No love was lost between the religious leadership of the proposed Utah Territory and the decidedly secular Congress in Washington, D.C. If the geographical barriers of the corner where Wyoming overlaps Utah had not been enough of a reason, Congress may have awarded it to Wyoming anyway —just to flex their will over the Mormon's attempt to assert their power.

To the south, Congress saw fit to establish a border in line with Colorado's southern border at the 37th parallel. They anticipated the work that lay ahead of them to divide the remaining land acquired after the Mexican War, and they had a firm commitment to keeping states as close in size to each other as possible. Setting Utah's southern border at the 37th parallel accomplished the mission of reining in Utah's desire for a larger state, while at the same time leaving room for the creation of two new states that would later be called Arizona and New Mexico.

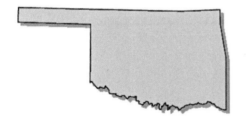

Chapter 46:
Oklahoma

Forty-sixth state, admitted November 16, 1907

Slavery was an extremely divisive issue in American history, and, as it turns out, that debate played a part in why Oklahoma has the long, narrow strip of land we call a panhandle.

Early in its evolution, the Republic of Texas stretched much farther north than it does today. But the Missouri Compromise of 1820 declared no state above the 36° 30' mark could be a slave-owning state. Faced with a dilemma, Texas made the decision to divest itself of a large chunk of its northern territory in order to continue the practice of slave ownership.

The territories that later became Colorado and Kansas absorbed the northernmost land Texas released, but an orphan strip was later created when Kansas' southern border was renegotiated. Congress elected to attach that strip to the territory of Oklahoma, which today we call the Oklahoma Panhandle.

With Kansas' renegotiated southern border came a firm establishment of Oklahoma's northern border at the 37th parallel.

Oklahoma territory spent a period of time divided into two entities. Maps created after 1890 showed "twin" territories referred to as Indian Territory and Oklahoma Territory. In 1893, the Dawes Commission embraced the task of uniting the two territories, but the standard political tactic at that time was to dissolve the Indian nations. Following accepted procedure, the Commission set out to grant American citizenship to the indigenous people, and incorporate their towns. But the leaders of The Five Civilized Tribes, as they called themselves, called their own commission to push to retain two separate states, and they named their state "Sequoyah." Congress never accepted Sequoyah as a state, and instead went forward with the ratification of Oklahoma as one state.

Oklahoma's southern border follows the Red River for its entire length. The designation appeared to have been neat and tidy, except the Red River has a northern branch and a southern branch at the western end of its run. Predictably Texas, Oklahoma's neighbor to the south, argued the northern branch should be the demarcation, making Texas' land holdings slightly larger. Oklahoma argued for the southern branch and, in 1897, the Supreme Court ruled in favor of Oklahoma's claim. Perhaps they believed Texas was big enough.

Unfortunately, the Red River demarcation between Oklahoma and Texas continues to cause issues on both sides of the border to this day. To read more about how the Red River boundary isn't settled, see the chapter "A River Runs Through It, and That's a

Problem" starting on page 27.

To the east, Oklahoma's border has an odd bow that bends into Arkansas. The original plan was to continue the straight line that was the bottom part of Missouri's western border, but a series of treaties with the Choctaw Indians resulted in the two segments that became the bowed boundary. The route to get there was ugly.

The intention of the original Treaty of Doak Stand was to remove the Choctaw Indians from Mississippi to what is now Oklahoma, but by mistake, the U.S. government assigned a larger piece of land to the Choctaw than they intended. When Congress "renegotiated" the treaty, they established a point 100 paces west of Fort Smith. From that point, one segment went south to the Red River and the other segment went north to the southwest corner of Missouri.

The word "renegotiate" implies that both sides came to an understanding. In reality, indigenous people were forcibly removed from land that had been granted them after they had already been violently removed from land elsewhere in the country. As white settlers moved west, the government made room for them by pushing out non-white landholders. The bow in the Oklahoma/Arkansas border is a visual reminder of an unpleasant show of might as a nation carved out land for its white settlers.

Chapter 47: New Mexico

Forty-seventh state, admitted January 6, 1912

A merica was built in stages, beginning with the thirteen original colonies. With westward expansion, the United States formed The Northwest Territory that later split into five states. As settlers moved across the Mississippi and the country bought the land known as the Louisiana Purchase, the size of the Union grew immensely. At different times, land traded hands between France, Spain and Britain. Acquisitions at the end of the Mexican War in 1848 were the source of the area that became New Mexico in 1912.

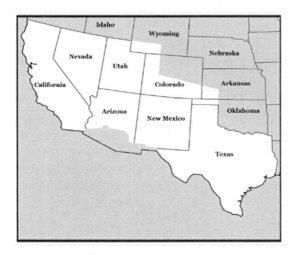

Before hostilities had even begun, the United States offered Mexico approximately double the amount they ended up settling for when the war ended. The U.S. paid $15 million (close to $500 million in today's money) for the land and also agreed to assume $3.25 million ($88 million today) in debts the Mexican government owed to U.S. citizens.

When Louisiana had been created, Congress' decisions on the locations of its borders had been influenced by the goal to keep a large French speaking population together. In the same vein, Congress produced a homogenous New Mexico by constructing borders that acknowledged its citizens were primarily Spanish speaking. Their language and cultural identity would be kept intact.

More than half of New Mexico's territory had once been a part of the Republic of Texas. When Texas ceded territory to the north in order to satisfy the Missouri Compromise (see chapter 28 on Texas for more information), they also sold their portion of the future New Mexico to the United States. Texas was deeply in debt, so the transaction proved a benefit to both Texas and the United States.

The decision on exactly where to locate New Mexico's eastern border showed the pre-planning Congress had exhibited when they created many of the other states. Size mattered, and the border on the 103rd longitude exposed a scenario in which two states, New Mexico and the future Arizona, which was ratified one month later, would be very close to the same size. Additionally, this decision would place the capital city of Santa Fe in the midpoint, east to west, of the new state. With this plan in mind, Congress established New Mexico's eastern border, and the western border took shape as well.

Colorado had become part of the Union 36 years earlier, and their southern border ultimately became New Mexico's northern border. But New Mexico's southern border required two steps to achieve the perimeter shown on modern maps.

As mentioned, Congress first acquired land for New Mexico from Mexico in 1848, but that purchase did not include the southernmost part of the state. In 1853, James Gadsden, president of the South Carolina Railroad Company, placed his stamp on the final shape of New Mexico. His appointment as the U.S. Minister to Mexico by President Franklin Pierce entrusted him with the task of buying one more piece of land from Mexico so that the U.S. could run a second transcontinental railroad through the southern part of the country.

Gadsden was successful. The "step-shaped" part of New Mexico's southern border became the territory necessary to allow important access to a passage for the much-needed railroad through the San Andreas Mountains.

Gadsden Purchase

Chapter 48:
Arizona

Forty-eighth state, admitted February 14, 1912

Arizona was the second state to benefit from the 1853 Gadsden Purchase as described in the chapter on New Mexico. Previous to that additional purchase of land from Mexico, Arizona had a southern border that stopped significantly farther north and was irregular in shape. The entirety of Arizona's land was acquired from Mexico, but it happened in two stages.

When the Mexican War ended in 1848, the United States purchased a vast amount of territory that looked like this:

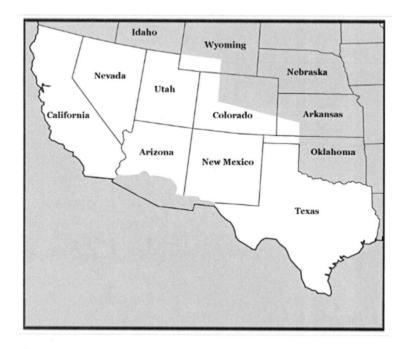

Five years later, James Gadsden added another piece. Gadsden had been hired by President Franklin Pierce to negotiate a transaction with Mexico to give the United States the land required to build a transcontinental railroad through the San Andreas Mountains.

Negotiating a treaty between parties that had just fought in a war is never easy. With Gadsden's leadership, both sides compromised, and the U.S. gained the land necessary to build a new railroad line. Mexico received cash to help stimulate its own economy after an expensive war.

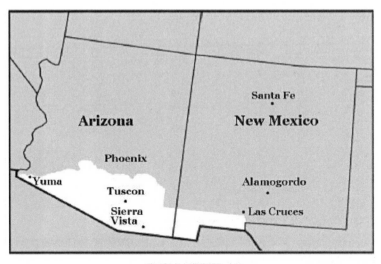

Gadsden Purchase

A close look at the new landmass reveals what appears to be a mistake. The new territory added to Arizona does not meet at the corner of California as one might expect. Gadsden had actually carefully crafted the line, knowing that by missing California's southwest corner by 20 miles, the new boundary would avoid the Gila Mountains. This border also carefully protects the town of Yuma, Arizona, and the critical flow of commerce along the Gila and Colorado rivers.

Arizona as a state was created when New Mexico divided its entire territory with a line that cut it in half vertically. Described throughout this book is Congress' intention to sculpt states of similar size whenever possible. At the time when New Mexico became a state in January of 1912, legislators immediately set the wheels in motion to create an equal-sized state of Arizona one month later. The

217

eastern border of Arizona is the western border of New Mexico, placed at the 109th meridian. In turn, this was the continuation of the line separating Colorado from Utah to the north. By using this demarcation, New Mexico and Arizona would be states of equal width.

Arizona's western border is the Colorado River, decided earlier when California joined the Union in 1850 and Nevada in 1864. Also in that design of Nevada, the decision to change the border to a straight line north—at the point where the Colorado River crosses the longitude of the Nevada/Utah border—completes the look of Arizona's western border.

Arizona was the final state to be sculpted from the North American landmass called the contiguous United States. Alaska and Hawaii wouldn't become states for another 47 years.

Chapter 49: Alaska

Forty-ninth state, admitted January 3, 1959

Unlike most of the lower 48 states, very few border decisions were necessary when the United States bought Alaska in the 1867 transaction that came to be known as "Seward's Folly." The shape of the territory carried over from the property Russia owned, including a straight-line demarcation between Russia and Canada at the 141st meridian. But why is there a long, narrow strip of land that continues to snake several thousand miles south along the coast of Canada?

This region, called the Alaskan Panhandle, was a stretch of land Russia also laid claim to when they owned the territory. When U.S. Secretary of State William Seward negotiated the purchase, he accepted Russia's map as evidence of exactly what he was buying. But Canada had a different perspective based on extremely vague wording of the original treaty with Russia. This disagreement had long been a bone of contention with Russia, and the dispute carried over to the United States when Seward made his purchase.

Finally, in 1903, the parties formed a tribunal to settle the debate. The panel consisted of three U.S. representatives, two Canadians, and one Brit. Unfortunately for Canada, their representatives expected the British judge to side with them, a British Dominion, against the Americans who fought against the British in a revolution. What they didn't count on was the importance Britain placed on regaining warmer relations with the United States at the beginning of the 20th century and how much they valued a strong ally in America.

The British judge sided with the U.S., but both sides eventually made compromises. In 1929, a Canadian scholar named Hugh Keenlyside wrote about the decision, calling the tribunal fair. Today, Canadian citizens who live in the northwestern part of British Columbia or the southern part of Yukon must pass through the United States to get to the coast.

For Americans, the panhandle provides the United States with a protected harbor between the Aleutian Islands and the coast of Canada. The land along the coast later became one of the favored routes to the Yukon Territory when prospectors discovered gold in that Canadian Province.

Why was the purchase of Alaska referred to as "Seward's Folly?" Critics believed the Secretary of State had spent an absurd amount of money ($7.2 million) on a piece of land that was mostly unexplored, and thus an unknown entity. In addition to the belittling nickname of "Seward's Folly," contemporaries also dubbed the transaction "Seward's Icebox" and "Andrew Johnson's Polar Bear Garden."

What had once been characterized as a "folly" at the time of Seward's purchase of Alaska quickly became a great asset because of the people, the discovery of gold and oil, and an abundance of other natural resources.

Chapter 50: Hawaii

Fiftieth state, admitted August 21, 1959

One would think in a discussion about how the states got their shapes, the reason for Hawaii's outline would be obvious. With a few pieces of land, each surrounded by the Pacific Ocean, the shape is defined without any contribution from humans.

But even Hawaii has a couple of surprises.

Because the islands of Hawaii originated from volcanic activity, the shape of the Hawaiian Islands continues to evolve slowly over time. The southern coastline of the "Big Island," for example, continues to change with recent dramatic lava flow. It is also important to recognize that erosion is a significant ongoing process for all of the islands, changing their footprint in the ocean forever. As of this writing, scientists at the University of Hawaii predict that Hawaiian beaches will erode by as much as 20 feet by 2050 and 40 feet by 2100, considerably altering the shape of Hawaii's islands.

In addition, the state of Hawaii isn't made up of only those eight islands frequently seen on maps. The Hawaiian Archipelago actually includes 18 islands (7 inhabitable) and 113 (small and uninhabitable) coral atolls that stretch across the Pacific Ocean for 1,500 miles. "What" you say?

In the 1800s ship captains discovered a long string of small, uninhabited islands in northern Polynesia. Many were perfect for what they needed: storage depots for coal. Steamships often traveled far from civilized lands, and they needed places to refuel. In time, they also discovered another valuable resource plentiful on the islands: bat guano. Also known as bat poop, this particular type of bat guano turned out to be an extremely valuable fertilizer, high in phosphorous and nitrogen and low in odor.

As in anything of value, this string of tiny uninhabited islands and reefs quickly drew interest from the U.S. government. Previous to this time, the United States did not have a colony or presence in the Pacific, and not only did these and other Pacific islands provide a convenient place for refueling for that vital Far East trade market, but the location in the Pacific gave the U.S. the ability to establish a naval base. At the time, a naval base had value in protecting the trade market, but military capability also served to be extremely important much later in World War II.

In the meantime, the royal rulers of Hawaii were expanding their own boundaries, and by the end of the 19th century claimed 16 islands, including the eight main ones recognizable on maps today. American industrialists (such as the Dole family) leased land from Hawaii for their businesses, and eventually managed to oust the royal family. In 1900, the U.S. declared Hawaii as a U.S. Territory, and by then it was comprised of the 1500-mile stretch of various landforms. Discussion began then about statehood, but it wasn't until 1959 when Hawaii was made the fiftieth state.

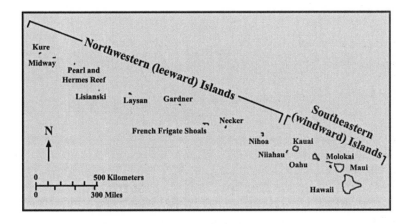

Glossary

Latitude and Longitude/Parallel
and Meridian: These terms describe the
way mankind has devised to indicate locations
on the globe.

Lines of **latitude** circle the earth horizontally and
lines of **longitude** are vertical, from pole to pole.
When expressed in degrees and minutes, the
designation describes the distance from the Prime
Meridian (Greenwich, England) for longitude, and
from the equator for latitude.

When a line is expressed as a **"parallel"** it refers to
the latitude because it is parallel to the equator. For
example, the border between the U.S. and Canada
from Minnesota to the Pacific Coast lies on the 49th
parallel.

When a line is referred to as a **"meridian"** it refers to
the longitude. Meridian is another term for the lines
that run from pole to pole.

When you are verbalizing the terminology of those
latitudes and longitudes, the symbol — ' — is read as
"minutes." So 36° 30' is read "36 degrees and 30
minutes north of the equator" or, alternately, "36 and
a half degrees north of the equator." See example on
following page.

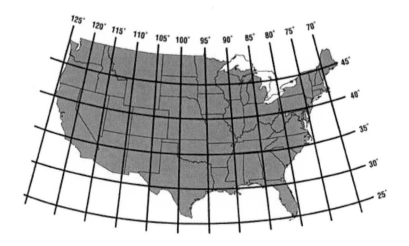

The United States: Are you wondering at what point Congress began using the term "United States?" What started as random settlements grew into colonies, then territories, and eventually became a union of states that is globally referred to as "The United States of America." The official naming happened on September 9, 1776 when the Continental Congress made its formal declaration. "United States" replaced the term "United Colonies," which had been in general use.

Panhandle: Any long, narrow, projecting strip of a state that is not a peninsula. This term is used for protruding parts of many states, including Texas, Alaska, Idaho, Florida and West Virginia, but the most visually obvious one is the panhandle of Oklahoma.

Bootheel: Just to clarify, old dictionaries seem to want this as two words. Newer dictionaries put it as one. I use the one-word version here. The only place it comes up is in the chapter about Missouri and the description seems quite apt.

Louisiana Purchase: This map shows the extent of land acquired in the Louisiana Purchase in 1803, later divided into the states indicated.

The Northwest Territory: This map shows the first intention to move west beyond the original colonies, later divided into the states indicated. The name derived from the fact that the territory was north and west of the Ohio River.

Northwest Territory

The Continental Divide: This term refers to the place on the North American continent where waters on each side flow to different oceans. There is more than one continental divide in the United States, but references here pertain to the "Great Divide" in the towering Rocky Mountains, where the natural feature influenced the placement of some of our states' borders.

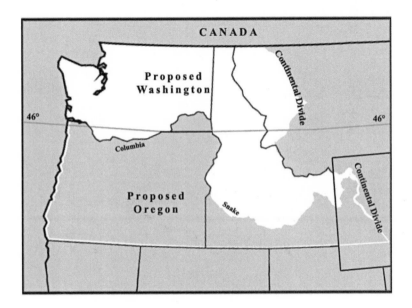

Bibliography

Stein, Mark, How the States Got Their Shapes, NY: Smithsonian Books, Harper Collins, 2008.

Linklater, Andro, Measuring America: How the United States was Shaped by the Greatest Land Sale in History, NY: The Penguin Group, A Plume Book, 2002.

Linklater, Andro, The Fabric of America: How our Borders and Boundaries Shaped the Country and Forged our National Identity, NY: Walker and Company, 2007.

Stein, Mark, How the States Got Their Shapes Too: The People Behind the Borderlines, Washington D.C: Smithsonian Books, 2011.

World Book Encyclopedia, 2014 Edition, Selected Volumes.

Wikipedia, Selected Volumes

Information on water disputes:

> http://www.legalgenealogist.com/blog/2012/0 7/23/as-the-river-flows/

> http://watchdog.org/247994/red-river-lawsuit-blm/